Deleuze and Science

Edited by John Marks

Printed and bound by CPI Antony Rowe, Eastbourne

Paragraph: A Journal of Modern Critical Theory

Volume 29 Number 2 July 2006

Special number: Deleuze and Science

Edited by John Marks

Contents

Introduction

JOHN MARKS

Since Gilles Deleuze's death in 1995, there has been much discussion of the eclectic and wide-ranging nature of his thought. A great deal has been written about his engagement with politics and social theory, as well as his highly influential work on art and literature. In recent times, attention has also been focused on Deleuze's interest in science. Deleuze's œuvre engages with work in the fields of mathematics, chemistry and biology, and refers to a number of key scientific writers, such as Gilbert Simondon, Ilya Prigogine and Isabelle Stengers, Jacques Monod and François Jacob. As far as the articulation of philosophy and science is concerned, Mark Bonta and John Protevi distinguish between Deleuze's project in his single-authored works, and those that are co-authored with Félix Guattari.[1] They argue that in his own work, Deleuze attempts to provide an ontology that corresponds to contemporary physics and mathematics. Deleuze and Guattari's co-authored works propose something slightly different, although obviously closely related, in the shape of an exploration of the usefulness of the contemporary biological and physical sciences for conceptualizing and acting in the world.

In the context of a body of work that is complex and challenging even for readers with a good grasp of philosophical ideas, this scientific material can often be quite forbidding. *Difference and Repetition*, for example, draws extensively on the mathematical fields of differential and integral calculus.[2] Similarly, *A Thousand Plateaus* draws on scientific concepts taken from physics, such as 'black holes', as well as mathematical ideas such 'fuzzy sets', 'neighbourhoods' and 'Riemannian spaces'. We might ask, then, just what draws Deleuze to these areas: what function do scientific theories and concepts fulfil in his work? We might also ask, as Sokal and Bricmont have done,[3] whether it is legitimate to wrench scientific concepts from their natural environment and put them to work in a very different, philosophical context. Such an undertaking runs the risk, for example, of using these concepts in a metaphorical way that could rob them of their coherence and relevance. Unsurprisingly, Deleuze was aware of these potential criticisms and provided a rationale for his own engagement with scientific ideas. When questioned about the use of scientific material in *A Thousand Plateaus*, he pointed out that the choices he and Félix Guattari made were influenced by a particular distinction that they made between two sorts of scientific notions:

> There are notions that are exact in nature, quantitative, defined by equations, and whose very meaning lies in their exactness: a philosopher or writer can use these only metaphorically, and that's quite wrong, because they belong to exact science. But there are also essentially inexact yet completely rigorous notions that scientists can't do without, which belong equally to scientists, philosophers, and artists. They have to be made rigorous in a way that's not directly scientific, so that when a scientist manages to do this he becomes a philosopher, an artist, too.[4]

It may be a case of alighting upon an 'inexact' notion in science, which Deleuze also refers to more precisely as 'an exact yet rigorous', or it may be a case of finding a particular component or potential within a scientific notion that philosophers and artists can legitimately work with.

In addition, as can be seen from the quotation above, Deleuze also proposes a way in which he thinks the fields of science, art and philosophy can, as it were, work off each other, without any single discipline exerting its superiority. The three fields have very different ways of functioning, all of which have their own internal validity and coherence, and it is a question of respecting that internal coherence. The task of science is to produce 'functions', whilst art seeks to produce sensory aggregates in the shape of 'affects' and 'percepts', and philosophy sets itself the task of constructing 'concepts'. The fact that these fields have their own distinctive ways of proceeding does not preclude interactions between functions, aggregates and concepts: all three areas can participate in the activity of thinking, or, in more precisely Deleuzian terms, *thinking difference*. He describes the relations of resonance and exchange that might emerge in terms of the interplay of separate melodic lines.[5]

Ontology, metaphysics, materialism

The focus of much of the recent work on the significance of science in Deleuze's thought has been upon the ways in which he finds in certain areas of science a support for his general project that is aimed towards an *ontology* of difference. It seems that the developing analysis of Deleuze's engagement with science has served to reinforce the notion that Deleuze's project is essentially ontological. Todd

May, for example, accords science an important position in Deleuze's construction of a creative ontology of difference, by means of which he seeks to explore the world of difference that 'both constitutes and disrupts' our tendency to view the world in terms of stable identities.[6] May suggests that Deleuze's philosophical reaction to this tendency to see the world in a relatively static way is slightly different to that of his contemporaries and fellow thinkers of difference, Foucault and Derrida. Although their work was similarly focused on the critique of 'sedimented', stable identities, Foucault and Derrida tended to emphasize the shifting ontological uncertainty upon which much of our conventional thinking is built. Deleuze, May indicates, goes a little further, in that he takes this critique of identity as a starting point for a new ontological project that takes difference on board. Deleuze himself alludes somewhat obliquely to this when he suggests that he was perhaps more naïve, more 'innocent' than other thinkers in his generation, in that he felt the least guilt about 'doing philosophy'.[7]

The construction of an ontology of difference aims to put us in contact, May argues, with the 'pure difference that forms the soil of all differences'.[8] This is the realm of the *virtual*, which is the key concept of Deleuze's ontology. The virtual should not be thought of as a field of *possibilities*, but rather as a very real realm of differences which constitute the *actual* world that we tend to perceive around us. In Deleuze's provocative and challenging ontology, the identity of an actualized object or event can never fully account for that object or event. This is where science enters into the picture. In straightforward terms, some areas of science may also have the capacity to put us into contact with, to 'palpate', as May puts it, difference: these are the areas of science that interest Deleuze most. (May points in particular the presence of Gilbert Simondon, Jacques Monod and Ilya Prigogine in Deleuze's work.) As we have seen already, scientific ideas can, as far as Deleuze is concerned, enter into a resonance with philosophy's task of bringing out the differences that disrupt many of our conventional ways of thinking and perceiving. What is more, these ideas can, in turn, resonate with artistic endeavours. Deleuze offers as an example of these types of resonances the way in which he found he could use the mathematical idea of the Riemannian space to think about cinema from the perspective of philosophy.[9] The 'functions', in scientific terms, of this space are, of course, rigorously defined. However, Deleuze extracts the general idea of neighbouring portions of space that can be reconfigured in a variety of ways and finds that it resonates with the way in which some post-war cinema — the films of Bresson for example — deals with space in challenging new ways. Similarly, Deleuze finds that Alain Resnais's innovative treatment of time resonates with the idea taken from physics, and outlined by Prigogine and Stengers, of the 'bakers transformation'.[10]

Rather than using the term 'ontology', Deleuze himself occasionally alludes to the fact that he sees his own work in terms of metaphysics. He describes himself as a 'pure metaphysician', and thinks of his work as an attempt, following Bergson, to provide modern science with the metaphysics that it lacks. Bonta and Protevi feel that Deleuze is successful in this aim, and they go so far as to claim that Deleuze might well be viewed as a contemporary equivalent of Kant, a claim echoed by Miguel de Beistegui in his recent work on Heidegger and Deleuze, *Truth and Genesis*.[11] That is to say, in the same way that Kant provided a philosophical system that corresponded to 'Euclidean space, Aristotelian time, and Newtonian physics', so Deleuze has provided a philosophical framework that corresponds to the contemporary scientific world of 'fragmented space', 'twisted time', and 'non-linear' physics.[12]

Recent interest in Deleuze's engagement with science has also brought out another, perhaps more accessible way, of considering these ontological/metaphysical preoccupations. This is the claim that Deleuze's work develops a new form of *materialism*. Deleuze finds, in thinkers like Spinoza and Simondon, a mode of thinking that challenges the dominant Western notion of matter as inert, passive 'stuff' that requires an external form in order to exist in a concrete and recognizable way in the world. Spinoza, for example, raises the possibility that matter might not require the external imposition of form; it might be that the capacity to take on a form is immanent to matter. That is to say, matter might be capable of *morphogenesis*. Manuel DeLanda sees, in works like *A Thousand Plateaus*, a new form of materialist philosophy being deployed.[13] In line with recent scientific developments in the fields of complexity and emergence, this 'neomaterialism' emphasizes the self-organizing properties of 'matter-energy'.[14] It is also encourages the view that the extensive material structures and objects that we see around us are simply 'coagulations or decelerations' of intensive flows of matter-energy that comprise the material world. Following Bergson, Deleuze and Guattari start from the assumption that matter is best thought of as energy, rather than as mass, and they emphasize what they see as the creative capacities of matter. This is the view of matter that Deleuze and Guattari elaborate in *A Thousand Plateaus*.

Here, they refer ultimately not simply to matter, but rather 'matter-movement', 'matter-energy' and 'matter-flow'. As opposed to the metric, extensive conception of matter conventionally associated with science, the focus is shifted to destratified, deterritorialized matter.[15]

Simondon and individuation

In order to grasp the fundamental ontological challenge that is at the heart of Deleuze's particular brand of materialism, it is necessary to consider briefly his encounter with the French philosopher of technology Gilbert Simondon. Deleuze draws on the work of Simondon in both *Difference and Repetition* and *A Thousand Plateaus*, where the latter's concept of *individuation* is incorporated into Deleuze's materialist ontology. As mentioned above, Simondon sets out to challenge the 'passive' conception of matter that he associates with conventional accounts. In particular, he finds the *hylomorphic* model unsatisfactory, in that it subsumes matter to form by virtue of the fact that it thinks in terms of fixed forms imposed upon homogeneous matter. In this way, the hylomorphic model effectively excludes crucial capacities of matter.[16] For one thing, as well as being the subject of forms, matter has its own 'energetic materiality in movement'; a notion that Deleuze and Guattari refer to in terms of implicit, *topological* forms that pertain to matter. Also, matter has what they call 'variable intensive affects'. Anybody who works with wood, for example, must recognize that it is more or less porous, elastic or resistant, and must 'surrender' to the wood, rather than attempting to impose a form upon it. Simondon aims to show that there is an intermediary dimension between form and matter which is, as Deleuze and Guattari put it, 'a space unto itself that deploys its materiality through matter'.[17] Rather than forms imposing themselves on matter, there are qualities in matter that forms bring out and facilitate. In short, the importance of Simondon's challenge to the hylomorphic schema for Deleuze and Guattari is to release matter from its passive, undynamic role and instead to emphasize that matter is in a state of continual flux and variation.

The critique of hylomorphism springs from Simondon's concept of individuation.[18] Simondon sets out to challenge the view that the individual living being fully accounts for the reality of that being. Conventionally, the individual has been viewed either as a 'self-centred' unity that defines its own essence — a substantialist approach — or, from a hylomorphic perspective, as the result of the conjunction of form and matter. This has led science to attempt to recreate the conditions that give rise to the individual. However, the problem with this way of thinking for Simondon is that we only look for factors that account for the individual as we see it now, in what we take as a sort of finished form: in short, an ontological privilege is accorded to the constituted individual. For Simondon, this approach cannot fully account for the *ontogenesis* — the developmental process — of the individual. Instead, Simondon proposes that we should think of the individual as having only a 'relative reality', since it constitutes only one phase of the 'whole being'. Rather than focusing on constituted individuals, we should think in terms of processes of *individuation*:

> *Thus, individuation is here considered to form only one part of an ontogenetic process in the development of the larger entity.* Individuation must therefore be thought of as a partial and relative resolution manifested in a system that contains latent potentials and harbors a certain incompatibility with itself, an incompatibility due at once to forces in tension as well as to the impossibility of interaction between terms of extremely disparate dimensions.[19]

Given that the individual is only a 'certain phase' of the whole being, Simondon suggests that individuation does not exhaust all the potentials that were contained in the preceding *preindividual* state. In other words, in order to approach the full reality of an individual living being, it must be seen in a wider context of ontogenesis; what is more, a wider context in which, in a sense, it differs from itself. Rather than viewing the living being as a stable, self-centred system, it should be thought of as a 'metastable' system that is in a process of *becoming*, a key concept in Deleuze's work.

As Todd May shows, Simondon's thinking on individuation is an important influence on the concepts of the virtual, the actual and *intensity* that Deleuze sets out initially in *Difference and Repetition*.[20] In turn, these concepts orientate Deleuze towards particular areas of, and developments in, science. Deleuze portrays the individuation of a living organism as the actualization of a virtual difference: as one 'solution' to a more general problem. As May indicates, the differential model of biology that Deleuze develops in this way connects with the moves that biologists themselves have made in recent times towards understanding life in terms of systems rather than individual beings. Current debates around DNA and the idea of a genetic 'code' might, for example, be conceptualized in terms of the distinction between the virtual and the actual. According to a widespread form of genetic essentialism, DNA is seen almost as being a pre-formed version of the future individual. Some would even go so far as to argue that this is the kind of thinking behind undertakings like the Human Genome Project. However, May suggests that we might usefully use a Deleuzian framework in order to think about genes in a different way, which would be very close to some of the current departures from the genetic determinism that has been so influential for the past thirty years or so:

> Think of a gene not as a set of discrete bits of information but instead as a virtual field of intensities that actualizes into specific concrete beings. The gene is not a closed system of pregiven information that issues out directly into individual characteristics. Instead, the genetic code is in constant interaction with a field of variables that in their intensive interaction generate a specific living being.[21]

Keith Ansell Pearson has recently shown how Deleuze uses this concept of individuation in order to recast Darwin and Weismann as thinkers of difference.[22] Deleuze focuses particularly on the way in which Darwin shows that the species is an 'illusion'; an actualization of the virtual field of constantly evolving individual differences.

Bergson

As well as Simondon's concept of individuation, any consideration of the position of science in Deleuze's work, as well as Deleuze's materialism, must deal with the ontology of becoming that he develops out of a sustained encounter with Bergson. In fact, there is a strong argument in favour of seeing Bergson as *the* key influence on a Deleuzian ontology/metaphysics. Constantin V. Boundas in particular has emphasized what he sees as the centrality of Bergson for Deleuze.[23] In Boundas' reading, Deleuze adopts a very particular ontology from Bergson, which has to do with continuous, intensive multiplicities: an ontology of *becoming*. This ontological view is perhaps best explained by looking at Bergson's critique of the 'natural' perception of movement. We do not fully grasp the real movement of an object if we think of that movement as being constituted by a series of instants, of static poses, as it were, that combine together to make movement. Instead, we must understand movement in a continuous sense. That is to say, in Deleuzian terms, movement must be thought of in terms of an *intensive continuum*, rather than in quantitative terms. The Deleuzian-Bergsonian world is in constant flux, and is not organized around a hierarchy of fixed reference points. It is a world of matter in flux that is composed of 'movement-images', which operate entirely independently of human consciousness. Human perception operates as a sort of screen that can capture only a portion of the intense luminosity that the world constantly generates.

In light of this, Boundas claims that Deleuze's work is motivated by a reading of the Bergsonian 'transcendental illusion', according to which we tend to focus on the *extensive* rather an *intensive* properties of the world around us. Boundas summarizes this illusion in the following way:

> It is the result of our exclusive preoccupation with discrete manifolds at the expense of continua, differences of degree at the expense of differences of nature, space at the expense of time, with things at the expense of processes, with solutions at the expense of problems, with sedimented culture at the expense of learning, with recognition at the expense of fundamental encounters, with results at the expense of tendencies. And as if such a list of allegedly similar errors were not enough, Deleuze sums it all up by saying that the transcendental illusion is the result of our exclusive preoccupation with the real and the possible at the expense of the virtual and the actual.[24]

This list of 'errors' constitutes an elegant summary of the ontological and ethical direction of Deleuze's entire project. The attempt — which draws on science — to shift the ontological focus to processes and differences of degree is inseparable from the drive to orientate ethics towards openness, availability and experimentation. For Bergson, 'analysis' focuses on the immobile dimensions of the world, whereas 'intuition' offers a route into a richer understanding of the mobility that characterizes the natural world.

'State' and 'nomad' science

Analysis is, in Deleuze and Guattari's terms, an approach that is sanctioned by what they call 'State' or 'Royal' science: it is at the heart of modern scientific and technological rationality. Intuition, on the other hand, tends to be dismissed as subjective and unreliable, but it fulfils for Bergson a role that is similar to the way in which 'nomad' or 'minor' science is conceptualized in *A Thousand Plateaus*.[25] Here, Deleuze and Guattari argue forcefully that the dominant tradition of State science has held in check a tradition of nomad science that engages in more subtle ways with the world of matter. Whereas analysis — the default mode of State science — immobilizes the world and extracts 'simples' from which reality can be reconstructed, intuition puts us in contact with the underlying continuity and fluidity of the natural world. Crucially, analysis neglects the dimension of temporality, attempting to extract repeatable structures from a world that is in constant flux.

Nomad science focuses upon the *expressive* and the *intensive*, as opposed to the *reductive* and *extensive* approach of State science. That is to say, whereas State science plots a closed space in which it places solid objects composed in a linear fashion, nomad science sets out an open space within which 'things-flow' are distributed. The open space of nomad science is, according to Deleuze and Guattari, 'vortical'. In spatial terms, the distinction between the two approaches is expressed as the difference between Euclidean and Archimedean geometry; between a *striated*, metric space and a *smooth*, topological space. Nomad science is based upon a 'hydraulic' model, in that it focuses on *flows* of energy and matter. It is an 'itinerant' form of science that follows the intensive states of systems in order to reveal virtual structures. In contrast, State science imposes a discipline on the eccentricities of nomad science, reformulating it in terms of 'civil and metric' rules. The State needs, both literally and metaphorically, to 'tame' the potential turbulence of hydraulic forces; to channel these forces by means of conduits, pipes and embankments. The difference between the construction of Romanesque and Gothic churches provides a further example. According to Deleuze and Guattari, Gothic architecture depends upon a fixed Euclidean model of form, which is then applied to matter. Romanesque architecture, on the other hand, is much more sensitive to the material (stone) with which it works.

As we have already seen, hylomorphism tends to treat matter as an inert, homogeneous substance, which requires form as a sort of 'mould' that structures it, whereas nomad science is more attuned to the 'singularities' of matter. For nomad science, everything 'is situated in an objective zone of fluctuation that is coextensive with reality itself'.[26] State science extracts laws and constants from the variations of matter, whilst nomad science is content to 'follow' the flow of matter. The notion of 'following', as opposed to State science's preference for 'reproducing', is key to an understanding of nomad science, and is related by Deleuze and Guattari to the activity of the journeyman. The State naturally seeks to organize and regulate the flow of labour, whereas the nomadic bodies of journeymen retained an autonomy and sympathy with their materials that was threatening to the organizational tendencies of the State. State science is, in short, highly effective at the necessary task of organizing matter and analysing a world of metric extension and linear causality. However, what it lacks, for Deleuze, is a metaphysical dimension that would enable it to *follow* the intensive contours of matter.

Complexity

One important contemporary expression of the analytic, extensive, 'State' tendency in science is the methodology of scientific reductionism. For the hard sciences reductionism involves understanding physical objects by looking at their component parts: the whole is considered to be an aggregate of these parts. 'Softer' sciences, such as the social sciences, have also sometimes adopted a reductionist methodology in order to analyse social phenomena. Of course, reductionism has long been challenged, and in recent years this challenge has been formulated in terms of the scientific concept of *complexity*. Complexity is not the same as 'complicated'. If a system is understood as complicated then, in principle, the claim is being made that a complete knowledge of this system is simply a matter of devoting enough resources to analyzing the complete structure and functioning of the system. However, if a system is understood as complex, then it is acknowledged that the complete structure and functioning of the system remains in some way unknowable and unpredictable. In short, complexity investigates the emergent, self-organizing, adaptive properties of systems.

As Isabelle Stengers points out, thinking on complexity is in some ways a response to a dominant model of scientific thinking that is based on an articulation of 'simplicity' and 'complication'.[27] She offers as an example of this model of scientific thinking Jacques Monod's claim, in *Chance and Necessity*, that the bacteria is the simple model that enables us, by extrapolation, to understand the genetic functioning of the rest of the living world.[28] The phenomenon of natural selection is the final piece in the jigsaw which provides a comprehensive description of the living world in terms of the simple and the complicated. However, complexity challenges a model of this sort by re-emphasizing the importance of the individual organism and its relationship with its environment. As Stengers indicates, the simple/complicated model that Monod outlines effectively reduces each living individual to the status of a 'superb reproductive automaton', carefully constructed by the blind but ultimately precise fine-tuning process of natural selection. (There is more than a hint of Dawkins' 'selfish gene' theory here.) Complexity, on the other hand, recognizes that many animals are, for one thing, capable of learning. Also, the interaction of the individual organism with its environment throughout the process of development means that phenotype cannot simply be read as a straightforward unfolding, or 'revelation', of genotype. Stengers proposes, therefore, that it might be more useful to think of evolution as a series of 'wagers': different species construct a more 'open' form of genetic development.

Bonta and Protevi have recently argued in some detail that Deleuze's work connects in various ways with complexity theory, otherwise known as 'nonlinear dynamics', and that such sciences suggest a world whose ontology can be mapped using Deleuze's three registers of the actual, the intensive, and the virtual.[29] They define complexity theory as the study of the 'self-organizing' capacities of systems through which energy and matter flows, otherwise known as 'open' systems.[30] Complexity theory has, in this way, moved biology in particular away from a preoccupation with the molecular components of living organisms, towards a focus on metabolism. That is to say, life is increasingly understood in terms of functional networks that are constituted by relationships between various processes: networks of genes, cells, organs and organisms. As Fritjof Capra puts it: 'The network is a pattern that is common to all life. Wherever we see life, we see networks.'[31]

Complexity theory is based on a nonlinear mathematical theory that is used to model the behaviour of nonlinear, nonequilibrium systems by constructing what are known as 'state spaces'. Following DeLanda, Bonta and Protevi identify three components which comprise the state-space of self-organizing systems. First, there are 'attractors', which model the main patterns of behaviour of the system. The attractor is a geometrical figure which represents the long-term tendencies of a system. Second, there are 'bifurcators', which constitute thresholds where a system 'flips' and changes its patterns of behaviour. Finally, there are 'symmetry-breaking events', which occur when 'bifurcators' come together and the system effectively transforms itself.

Complexity theory has had a particularly significant impact in the field of *morphology*, the study of biological form. The development of a predominantly mechanistic, reductionist understanding of genetics in the second half of the twentieth century meant that morphology was somewhat neglected until nonlinear dynamics revived the interest in functional networks.[32] One important result of this revival of interest in morphology has been to call into question the notion that living organisms are almost entirely determined by a genetic 'programme' or 'blueprint'. Instead, the focus has shifted to the influence of epigenetic, metabolic and environmental networks. In this way, complexity theory poses a direct challenge to reductionism, since it looks at the ways in which complex systems can demonstrate properties that cannot be explained in terms of an aggregate of their constituent parts. As Bonta and Protevi point out, complexity theory also enables us to think an alternative to the 'oscillation' between structure and agency that characterizes much thinking in the social sciences.[33] Deleuze and Guattari's commitment to materialism means that they eschew the rigidities of linguistic structuralism in favour of analyses of the way in which systems — biological, physical, social — demonstrate *self-organizing* tendencies that cannot be traced back to the agency of specific individuals or components within the system. Deleuze and Guattari seem to be inspired here by complexity theory's claim that physical and biological systems can interact with their environment and trigger 'self-organizing' processes. In this way, Bonta and Protevi argue, *Anti-Œdipus* and *A Thousand Plateaus* can be read through complexity theory, in order to escape the twin conceptual blockages of seeing 'structure' as either a closed homeostatic system or a chain of linguistic signifiers, and 'agency' as a enigmatic, vital quality inherent to individuals.

Intensive Science

In recent times, Manuel DeLanda has brought together much of this material in a sustained attempt to show that there are forms of 'intensive' science that would appear to incorporate some of the dimensions that Deleuze and Guattari feel are missing from State science. Delanda has, in short, proposed an intensive ontology of the material world that is directly inspired by Deleuze (and Guattari). This intensive ontology has its roots in DeLanda's reading of Deleuze's materialism.[34] DeLanda initially takes Deleuze and Guattari's work as a significant influence in his analysis of a contemporary paradigm shift in the direction of an understanding of life in 'nonorganic' terms. One of the key insights of this paradigm shift is to show that both living and non-organic systems depend upon 'intense' flows of matter and energy. He points, for example, to the concept of the *body without organs* (BwO) in *A Thousand Plateaus* as a formulation that corresponds effectively to the degree-zero, as it were, of flows of matter-energy that are unformed and destratified. As Deleuze and Guattari indicate, the BwO in this broad sense is composed uniquely of intensities.

This immanent field of intensities operates as a flow from which recognizable, stratified forms emerge in the world around us.

One of the first ways in which DeLanda develops this 'inhuman' neomaterialist perspective is by adopting a historical perspective based upon the *longue durée*: a 'geological philosophy', as he calls it. So, for example, he suggests that the earth's crust is, in terms of the nonlinear dynamics of the planet, simply a 'hardening' that emerges from a system of underground lava flows.[35] A geological time frame — what we might think of as a geological *durée* in Bergsonian terms — reveals that structures such as mountains are created by underground lava flows and the forces they exert on plate tectonics. In this sense, geological structures such as mountains may appear to be permanent and durable structures, but from another perspective they constitute a 'slowing-down' of global flows of matter-energy. A similar perceptual operation can be performed with respect to individual human minds and bodies. In historical and populational terms, individuals are simply temporary coagulations in the flow of biomass, genes and memes (units of 'culture') that move through time across the world. In the light of conventional geological science, and in an era of time-lapse photography, most of us would have little problem in imagining the sort of long-term processes that create mountains. Similarly, basic concepts from modern genetics are sufficiently widely circulated for the notion of the flow of genes across generations to be relatively unremarkable. However, the concept of the body-without-organs as an immanent field of intensities is as extraordinary as the Bergsonian metaphysics of movement-images outlined above. Nonetheless, DeLanda proposes that we should, as far as it is possible, take this philosophical physics as being literal rather than metaphorical. That is to say, we should construct a fully-fledged ontology out of these insights, and in order to do this he engages in the project of drawing systematic connections between Deleuze and Guattari's work and current scientific theories. This is the project that DeLanda undertakes in *Intensive Science and Virtual Philosophy*, where he explores in detail the ways in which emergence, non-linear dynamics and complexity theory complement and correspond to Deleuze's 'virtual' philosophy.

DeLanda makes the claim that Deleuze's realist ontology is comprised of three distinct ontological dimensions: the virtual, the intensive and the actual. In light of this, before looking at what DeLanda means by intensive science, it is worth rehearsing briefly Deleuze's concepts of the actual and the virtual. For Deleuze, there is no purely actual object, in the sense that the actual and the virtual are inseparable. In more concrete terms, the interaction between the actual and the virtual that Deleuze proposes means that the actual form of something does not define the ways in which that thing may differ or change. As he emphasizes in *Difference and Repetition*, the virtual is, in this sense, not simply a realm of possibilities, but rather another dimension of the real.[36] The virtual is not like the possible, in that it is simply waiting to be realized, since it is already fully real. Instead, it needs to be actualized, and this entails a process of what Deleuze calls 'differenciation'.

In *Intensive Science and Virtual Philosophy*, DeLanda shows how Deleuze's virtual ontology is opposed to any form of essentialism, and how contemporary developments in mathematics and science point in a similar direction. In mathematical terms, DeLanda suggests that 'topological' geometry, which is concerned with the properties of geometric figures that remain invariant even when they are bent, stretched or deformed, provides a metaphorical route into understanding how metric, actual space is 'born', as he puts it, from a topological continuum. In *A Thousand Plateaus*, he points out, Deleuze and Guattari explore just such a materiality, or corporeality as they also call it, which proposes an alternative to essentialist or typological perspectives:

> This corporeality has two characteristics: on the one hand, it is inseparable from passages to the limit as changes of state, from processes of deformation or transformation that operate in a space-time itself an exact and that act in the manner of events (ablation, adjunction, projection...); on the other hand, it is inseparable from expressive or intensive qualities, which can be higher or lower in degree, and are produced in the manner of variable affects (resistance, hardness, weight, color...).[37]

DeLanda points to the fact that Deleuze deploys the concept of multiplicity in order to shift the focus of his work away from *essentialism* of all kinds. Instead he focuses on the *morphogenetic* processes that are defined in terms of the dynamics of matter and energy, as opposed to essentialism's frequent recourse to abstract categories such as types or ideal forms. DeLanda indicates that the term 'multiplicity' is closely related to the mathematical term 'manifold'.[38] 'Manifold' belongs to the tradition of differential geometry associated with Friedrich Gauss and Bernhard Riemann. According to DeLanda, Gauss and Riemann develop the manifold in order to explore the concept that a surface may be a space 'in itself'.

In summary, in *Intensive Science and Virtual Philosophy* DeLanda gives a number of examples in the physical and biological realms of how the virtual is actualized by means of intensive processes. The flat ontology that emerges from this encounter with Deleuze has much in common with the ontology of Deleuze-Bergson discussed above. The world that DeLanda portrays is composed of intensive and differential dimensions and is consequently in a continual state of flux and transformation: it is world without essences and transcendent factors, in which matter displays self-organizing and emergent properties.

Deleuze and Science

The papers collected in this volume explore a number of the issues raised in the preceding brief discussion of the functioning of science in the work of Deleuze-Guattari. The papers also investigate the connections between this work and current developments in the fields of science and mathematics. John Protevi examines the connections between Deleuze and Guattari and recent developments in the field of complexity theory, and looks in particular at the concept of emergence. His paper explores the usefulness of Deleuze and Guattari in understanding the way in which complexity theory and emergence contribute to current debates on the nature of reductionism in the fields of biology and the social sciences. In the course of this discussion, Protevi sets out in some detail what he sees as the connections between Deleuze and Guattari and the concept of self-organization, as put forward by Francisco Varela.

Coming from a background in quantum physics, David Holdsworth engages directly with DeLanda's reading of Deleuze's ontology of the virtual. As well as exploring the mathematical aspects of DeLanda's analysis, Holdsworth also alludes to literary theory in order to defend experimentation with conceptual resonances across disciplines as disparate as mathematical physics and literary criticism. In a similar vein, Arkady Plotnitksy explores the connections between quantum field theory and Deleuze's concept of the virtual. Plotnitsky draws in particular on the idea that science and philosophy have a shared interest in constructing forms of thought that confront *chaos*. Simon Duffy also focuses on Deleuze and Guattari's understanding of the relationship between science and philosophy in order to revisit the intellectual disagreement that arose between Benedict de Spinoza and the British scientist Robert Boyle in the seventeenth century. In contrast to readings that tend to polarize the debate, Duffy suggests that Deleuze and Guattari's thinking on the relationship between science and philosophy provides a more productive reading of the dialogue between Spinoza and Boyle. In this way, he suggests, the so-called 'Spinoza-Boyle' controversy in turn serves to illuminate Deleuze and Guattari's exploration of the differences between science and philosophy.

Whilst he accepts the explanatory force of DeLanda's highly influential recasting of Deleuze's ontology, James Williams takes issue with certain aspects of this reading. He feels that DeLanda's attempt to forge such a close series of connections between Deleuze's work and a particular set of contemporary scientific theories does not do full justice to the ontological openness of this work. What is more, he suggests that there is a metaphysics at the heart of Deleuze's work that cannot be entirely accounted for by means of such a rigorously defined 'scientific' approach. He develops his argument through a discussion of the differences between Deleuze's and Bachelard's very different philosophies of science.

John Marks and Matthew Kearnes look at the ways in which Deleuze's work can help us to gain critical purchase on key recent developments in the fields of science and technology. With reference to debates on reductionism and genetic determinism, complexity theory, and DeLanda's concept of intensive science, Marks examines Deleuze and Guattari's somewhat enigmatic but highly significant encounter with molecular biology. Marks focuses in particular on references to the French molecular biologists François Jacob and Jacques Monod in order to argue that Deleuze rather presciently draws out intensive potentials that have, until recently, been left unexplored in a field of science that has often been conceptualized in explicitly reductionist terms. Kearnes, on the other hand, focuses on the cutting-edge field of nanotechnology. Taking as his starting point Deleuze's concept of 'control societies', Kearnes argues that Deleuze formulates a philosophy of technology that is particularly open to internal variation, and which therefore serves as a powerful tool for understanding the functioning of nanotechnology. He also suggests that Deleuze's ontology of difference can be deployed to counter some of the more hubristic claims associated with new forms of technology of this type.

Taken together, the contributions to this volume provide a series of starting points for discussion and reflection in a field where much remains to be explored.

NOTES

1 Mark Bonta and John Protevi, *Deleuze and Geophilosophy: A Guide and Glossary* (Edinburgh, Edinburgh University Press, 2004), 12.
2 Gilles Deleuze, *Difference and Repetition*, translated by Paul Patton (London, Athlone, 1994).
3 See Alan Sokal and Jean Bricmont, *Intellectual Impostures: Postmodern Philosophers' Abuse of Science* (London, Profile, 1998).
4 Gilles Deleuze, *Negotiations*, translated by Martin Joughin (New York, Columbia University Press, 1995), 29.
5 Deleuze, *Negotiations*, 125.
6 Todd May, *Gilles Deleuze: An Introduction* (Cambridge, Cambridge University Press), 19.
7 Deleuze, *Negotiations*, 89.
8 May, *Gilles Deleuze: An Introduction*, 21.
9 Deleuze, *Negotiations*, 124.
10 *Negotiations*, 124.
11 Miguel de Beistegui, *Truth and Genesis* (Bloomington, Indiana University Press, 2004).
12 Bonta and Protevi, *Deleuze and Geophilosophy*, vii–viii.
13 See Manuel DeLanda, *A Thousand Years of Nonlinear History* (New York, Zone Books, 1997) and *Intensive Science and Virtual Philosophy* (London, Continuum, 2002).
14 On matter and energy in Deleuze's ontology see Constantin V. Boundas, 'Deleuze-Bergson: an Ontology of the Virtual', in *Deleuze: A Critical Reader*, edited by Paul Patton (Oxford, Blackwell, 1996), 96–7.
15 Gilles Deleuze and Félix Guattari, *A Thousand Plateaus: Capitalism and Schizophrenia*, translated by Brian Massumi (London/Minneapolis, University of Minnesota Press, 1987), 407.
16 Deleuze and Guattari, *A Thousand Plateaus*, 408.
17 *A Thousand Plateaus*, 409.
18 See Gilbert Simondon, 'The Genesis of the Individual', translated by Mark Cohen and Sanford Kwinter, in *Zone 6: Incorporations*, edited by Sanford Kwinter and J. McCrary (New York, Zone Books, 1992), 297–319. This is a translation of the introduction to Gilbert Simondon, *L'individu et sa genèse physico-biologique* (Paris, Presses Universitaires de France, 1964).
19 Simondon, 'The Genesis of the Individual', 300.
20 Deleuze, *Difference and Repetition*, 88.
21 May, *Gilles Deleuze: An Introduction*, 88.
22 Keith Ansell Pearson, *Germinal Life: The Difference and Repetition of Deleuze and Guattari* (London, Routledge, 1999), 90–2.
23 Boundas, 'Deleuze-Bergson', 81–106.
24 'Deleuze-Bergson', 85–6.

25 Deleuze and Guattari, *A Thousand Plateaus*, 361–74.
26 *A Thousand Plateaus*, 373.
27 Isabelle Stengers, *Power and Invention: Situating Science*, translated by Paul Bains (London/Minneapolis, University of Minnesota Press, 1997), 13.
28 Jacques Monod, *Chance and Necessity: An Essay on the Natural Philosophy of Modern Biology*, translated by Austryn Wainhouse (London, Collins, 1972).
29 Bonta and Protevi, *Deleuze and Geophilosophy*, 16–21.
30 *Deleuze and Geophilosophy*, 17.
31 Fritjof Capra, 'Complexity and Life', *Theory, Culture & Society*, 22:5 (2005), 34.
32 Capra, 'Complexity and Life', 39.
33 Bonta and Protevi, *Deleuze and Geophilosophy*, 3.
34 Manuel DeLanda, 'Immanence and Transcendence in the Genesis of Form', in *A Deleuzian Century*, edited by Ian Buchanan (Durham and London, Duke University Press, 199), 119–34.
35 DeLanda, *A Thousand Years of Nonlinear History*, 257.
36 Deleuze, *Difference and Repetition*, 208–14.
37 Deleuze and Guattari, *A Thousand Plateaus*, 407–8.
38 DeLanda, *A Thousand Years of Nonlinear History*, 10.

Deleuze, Guattari and Emergence

JOHN PROTEVI

Abstract
The concept of emergence — which I define as the (diachronic) construction of functional structures in complex systems that achieve a (synchronic) focus of systematic behaviour as they constrain the behaviour of individual components — plays a crucial role in debates in philosophical reflection on science as a whole (the question of reductionism) as well as in the fields of biology (the status of the organism), social science (the practical subject) and cognitive science (the cognitive subject). In this article I examine how the philosophy of Deleuze and that of Deleuze and Guattari can help us see some of the most important implications of these debates.

Keywords: Deleuze, Guattari, emergence, complexity, materialism, Varela

Emergence

The concept of emergence — which I define as the (diachronic) construction of functional structures in complex systems that achieve a (synchronic) focus of systematic behaviour as they constrain the behaviour of individual components — plays a crucial role in debates in philosophical reflection on science as a whole (the question of reductionism) as well as in the fields of biology (the status of the organism), social science (the practical subject), and cognitive science (the cognitive subject).[1] In this article I examine how the philosophy of Deleuze and that of Deleuze and Guattari[2] can help us see some of the most important implications of the debate on the status of the organism, as well as prepare the ground for a discussion of the practical and cognitive subject.

All of what follows depends on accepting the strong case put forth in Manuel DeLanda's *Intensive Science and Virtual Philosophy* that Deleuze's project in *Difference and Repetition* and *The Logic of Sense* — continued in the collaborative works of DG — establishes the ontology of a world able to yield the results forthcoming in complexity theory.[3] In terms I will explain further below, complexity theory models material systems using the techniques of nonlinear dynamics, which, by means of showing the topological features of manifolds (the distribution of 'singularities') affecting a series of trajectories in a phase space, reveals the patterns (shown by 'attractors' in the models), thresholds ('bifurcators' in the models), and the necessary intensity of triggers (events that move systems to a threshold activating a pattern) of these systems.[4] By showing the spontaneous appearance of indicators of patterns and thresholds in the models of the behaviour of complex systems, complexity theory enables us to think material systems in terms of their powers of immanent self-organization.

There are four main benefits here. (1) The first is the critique of hylomorphism, that is, the notion that matter is chaotic or passive and so in need of rescue (by means of the laws of God, or a transcendental subject, or the scientific project) to provide it with order or novelty.[5] (2) We can thus avoid the issue of reduction to physics, the science whose laws predict the behaviour of 'matter' at its simplest.[6] (3) Furthermore, by modelling the negative and positive feedback mechanisms characteristic of complex systems, complexity theory thereby enables us to ground the concept of emergence in the effects of such mechanisms.[7] (4) And as a last benefit, complexity theory enables us to dispense with the false problem of 'downward causation' by showing that the constraints of a pattern, described by an attractor, are not a case of efficient causality, but instead need to be thought of as a 'quasi-cause'[8] (where quasi-causes are said to replace final causes) or a reformed formal and final cause.[9]

Difference and Repetition will thus enable us to talk about synchronic emergence (order) and diachronic emergence (novelty), but it is the collaborative works of the emergent unity, DG, which enable us to talk about a third kind of emergence by letting us situate the organism or subject as one emergent unity in a field of such unities. Such a field is not a simple hierarchy of levels, for besides allowing us to move 'below' to modules or agents from which the organism or subject emerges and 'above' to social machines in which the organism or subject is a component of an emergent unity, DG also enable us to move 'transversally' to assemblages formed from biological, social and technical components. This third form of emergence, transverse emergence in assemblages, is what I call 'political physiology'.

Complexity theory

A first distinction: complexity theory is not chaos theory. Chaos theory treats the growth of unpredictable behaviour from simple rules in deterministic nonlinear dynamical systems, while complexity theory treats the emergence of relatively simple functional structures from complex interchanges of the component parts of a system. Chaos theory moves from simple to complex while complexity theory moves from complex to simple.

To explain how complexity theory studies the emergence of functional structures we need to understand three sets of linked concepts: (1) in the system being modelled: range of behaviour, fluctuation, patterns and thresholds; (2) in the dynamical model: phase space, trajectory, attractors and bifurcators; (3) in the mathematics used to construct the model: manifold, function and singularity. A phase space is an imaginary space with as many dimensions as 'interesting' variables of a system; the choice of variables obviously depends on the interests of the modeller. The phase space model is constructed using a manifold, an n-dimensional mathematical object. The manifold qua phase space represents the range of behaviour open to the system: 'what a body can do'. The global condition of a system at any one point can be represented by a point in phase space with as many values as dimensions or 'degrees of freedom', to use complexity theory jargon. If you track the system across time, you can see the point trace a trajectory through the manifold/phase space, a trajectory representing the behaviour of the system. For some systems you can solve the equation that governs the function represented by that trajectory; for other systems you must simply run a computer simulation of the model and see what happens. Often these simulations will show the trajectories following a particular configuration. These shapes of trajectories are called 'attractors' and represent patterns of behaviour of the real system. There are various kinds of attractors: point (stable or steady-state systems), loop (oscillating systems), and 'strange' or 'fractal' (turbulent or 'chaotic' systems). Here we must make an important distinction: chaos in the sense of chaos theory is not the ancient cosmogonic sense of chaos, which is now called a 'random' system, one whose model produces no attractors. On the contrary, the models of what are now called chaotic systems do have attractors, albeit fractal ones. Although the behaviour of chaotic systems is unpredictable in quantitative detail it is sometimes predictable in the long run or 'qualitatively'.

Now the areas of phase space surrounding attractors — representing the normal behaviour of the system in one or another of its behaviour patterns — are called 'basins of attraction'. The behaviour patterns described by attractors are formed by the action of negative feedback mechanisms. The layout of attractors in the phase space, which describes the layout of the patterns of behaviour of the system, is defined by the layout of singularities, which are mathematical objects that define the topological structure of the manifold; a singularity is a point where the graph of the function changes direction as it reaches local minima or maxima, or more dramatically, where the slope of the tangent to the graph of the function becomes zero or infinite. A singularity in the manifold indicates a bifurcator in the phase space model which in turn represents a threshold where the real system changes qualitatively. A singularity is not an attractor, but it defines where attractors are found by indicating the limits of basins of attraction. Some bifurcators are 'thick', so that inside a zone of sensitivity we find that minimal fluctuations — those that would otherwise be swallowed up by the negative feedback loops whose operation defines the normal functioning of a system following one of its patterns of behaviour — can now push the system to a new pattern. In model terms, in zones of sensitivity or crisis situations we find fractal borders between basins of attraction, so that any move, no matter how small and in no matter what any direction, might — or might not — trigger the move to another basin of attraction. Here we have an irreducible element of 'chance', even though the system is thoroughly deterministic.

As we have said, what keeps a system inside a behaviour pattern — represented by the trajectories inhabiting a basin of attraction — is the operation of negative feedback loops that respond to system fluctuations below a certain threshold of recuperation by quickly returning to the system to its pattern. (These fluctuations can be either endogenously generated or act as responses to external events.) These quickly recuperating systems are called 'stable'. With regard to normal functioning, fluctuations are mere 'perturbations' to be corrected for in a stable system. Since internal system resources translate the sense of events into terms significant to that system, external events are merely 'triggers': they trigger a pre-patterned response.[10] Such changes in environment relevant to the system's 'interests' are called 'signs'. (From this perspective, which Deleuze and Guattari buy into, signs therefore extend far beyond human language and meaning is the probability of triggering a response in a system. Thus Deleuze and Guattari are not 'postmodernist' in the sense that they are primarily or exclusively concerned with meaning as produced in chains of signifiers, because for them, 'signs' are not restricted to 'signifiers'.)[11]

Now fluctuations of a certain magnitude — beyond the recuperative power of the negative feedback loops or homeostatic mechanisms — will push the system past a threshold, perhaps to another pattern in its fixed repertoire, or perhaps into a 'death zone' where there are no patterns but only static or chaos. Thus some stable systems are 'brittle': they can be broken and die. Some systems are 'resilient' however: a sign or trigger that provokes a response that overwhelms its stereotyped defensive patterns and pushes the system beyond the thresholds of its comfort zones will not result in death but in the creation of new attractors representing new behaviours. We call this 'learning'. (Although of course there is a sense in which the old system has died and the new one is 'born again'. All sorts of questions of personal identity could be raised here.) Sometimes this learning, this creation of new patterns for a particular system, repeats patterns typical of systems of its kind; we call this 'development'. Sometimes, however, this learning is truly creation: we call this 'evolution', or as we will see, 'diachronic emergence'.

Diachronic emergence, or creativity in the production of new patterns and thresholds of behaviour, is what Deleuze will call an 'event', which is not to be confused with a mere switch between already established patterns or with the trigger or 'external event' that pushes the system past a threshold and produces the switch. The Deleuzian event repatterns a system. The key to the interpretation of Deleuze in DeLanda's *Intensive Science* is that the virtual is the realm of patterns and thresholds, that is, those multiplicities, Ideas, or abstract machines that structure the intensive morphogenetic processes that produce actual systems and their behaviours. A behaviour pattern, or a threshold at which a behaviour pattern is triggered, needs to be ontologically distinguished (or 'modally' distinguished) from behaviour, just as singularities are distinguished from ordinary points on the graph of the function.[12] Thus patterns and thresholds are virtual, while behaviour is actual. An event, in creating new patterns and thresholds, restructures the virtual.

Let me concretize the discussion so far with the example of weather. Weather is a classic chaotic system: Lorenz, the butterfly effect and so on.[13] But as Waldner explains, you have to remember that while weather is chaotic, there are definite long-term patterns; weather is unpredictable, but climate is not.[14] So while the butterfly's flap might move a developing weather system to another point in its sensitive zone (represented by a move to a point on its chaotic attractor minimally close by on a fractal border), and while this fluctuation might perhaps even switch that particular weather system to a hurricane pattern, it will not budge the global climate system out of its pattern, for there are a predictable number of hurricanes per season — on average, over the long duration — and here butterflies have no effects. Of course, the nightmare scenario of nuclear winter is that the effects of large-scale thermonuclear explosions will flip the global climatic system into its other big pattern: the ice age. Has global warming already pushed us into a crisis situation, modelled by a zone of sensitivity in which a minor fluctuation that otherwise would not have budged us off our attractor, but merely moved us to another point on that attractor, will now push us into another attractor, the ice age attractor, that is, into a new climate pattern? Or will the global system create a new pattern, neither temperate nor ice age, but something different? The global climate system might be creative and resilient, but there is no guarantee that the new pattern will provide a viable environment for human beings!

Synchronic emergence

Now that we have discussed the basics of complexity theory, which will provide the basis for the rest of our discussion, let us discuss synchronic emergence. Diachronic emergence, as we have intimated, is the creation of new patterns and thresholds in a system. Synchronic emergence has, however, dominated the discussion so far. Unfortunately. First, the definition: a synchronically emergent structure is that which enables focused systematic behaviour through constraining the action of component parts. This definition encapsulates what Thompson and Varela call 'reciprocal causality': the mutual constitution of local-to-global or 'upward' causality that produces focused systematic behaviour and the global-to-local or 'downward' causality that constrains the local interactions of components.[15] Synchronic emergence is the emergence of 'order out of chaos', as the popular formula has it.[16]

The focus on the part/whole relation of synchronic emergence has caused a lot of mischief in social science with the structure/agency dilemma, and in the philosophy of mind with the entire range of problems surrounding the issues of physicalism, eliminative materialism, reductionism, supervenience, and so forth.[17] We see a curious chiasmatic relation here. In consciousness issues, researchers operating without a notion of complex systems struggle to relate the global level of freedom (the mental whole) to the local determinism of physical parts, while in social science they struggle to relate the local freedom of individual agents (parts) to the global determinism of social structure (the whole). The relation of methodological individualism in social science to genetic reductionism in biology is not chiasmatic however, but analogic. Genetic reductionism is analogous to methodological individualism in that all living or social phenomena are considered mere epi-phenomena of fundamental units (genes or agents); in other words, these stances accept only 'upward causality'.[18]

How is one to demonstrate the existence of a synchronic emergent functional structure rather than just asserting its presence? A negative demonstration would be to point out any systematic behaviour that cannot be accounted for by analysis of the system into its components and then explaining the behaviour of the whole on the basis of the properties of the parts. But this explanatory failure might simply be epistemic (due to sensitivity to initial conditions and so on), and so what we have here is merely 'epistemological emergence'; we are not any closer to being able to claim 'ontological emergence' or emergence as a real feature of the world. Thus we see that a key arena for emergence questions is the unity of the sciences issue. Emergentists will propose level-specific laws, while reductionists will claim them to be merely 'epistemological emergence' or simply markers of our (temporary) ignorance. Kant struggled with this problem in the *Critique of Judgment* when he relegated teleological judgments to the realm of regulative principles.[19]

Now what used to be called ontological emergence can sometimes be shown to have been merely epistemological emergence due to the immaturity of science at the time. For instance, Mill proposed water as emergent: H_2O does not act like the 'combination' of hydrogen and oxygen. But quantum mechanics has shown ways to explain water's properties on the basis of the properties of hydrogen and oxygen.[20] Note also that systemic focus as the criterion for emergence is not the same as what some people call the 'emergence' displayed by thermodynamic properties. A single molecule does not have a temperature but temperature only 'emerges' at the level of the whole, as average kinetic energy. But there is nothing systemic here: there is no notion of 'normal function' or 'breakdown' applicable to thermodynamic phenomena.[21]

This brings us to the notion of arguing for ontological emergence by demonstrating the formation of attractors in computer modelling of systems, since such attractors represent the functioning of nonlinear feedback loops.[22] Such attractors would entail a reduction in the amount of the state space available to the system, and hence indicate the possibility of focused behaviour achieved through the constraint of components. We would dispense thereby with the problem of 'downward causation' when that is thought of as efficient causation issuing from a reified totality, as in the controversies surrounding 'mental causation' and sociological realism.[23]

Part of the difficulty in these controversies is a widespread misunderstanding of efficient and final causes in Aristotle's *Physics* 2.3, where, notoriously, Aristotle uses the poietic model of statue construction to explain the biological phenomenon of development. What we have come to call the efficient cause is that 'from which' the development occurs: the commanding origin of motion/development, the sculptor or father. It is not the hammer blows or the immediate action of the spermatic fluid; it is not billiard ball causality. The final cause is the perfect state of development or end state of motion. While the sculptor might have something in mind to guide his work, we need not impute purpose or intention to biological development. The final cause or end state channels development; the infant does not intend to grow into an adult. It is this notion of channelling which is the key to understanding systematic constraint and focused behaviour in synchronic emergent functional structures. In other words, synchronic emergence is a misnomer; there is always a coming into being of functional structures which needs to be conceptualized.

Synchronic emergence works on one level by means of negative feedback mechanisms. Now a negative feedback mechanism is just that, a mechanism. It can be made sense of on the billiard ball model of efficient causality. These actions produce constraints on behaviour, producing patterns of behaviour modelled by basins of attraction, and these constraints produce focused system-wide behaviour. The problem is how to conceptualize the onset of such focused behaviour, modelled by the funnelling of trajectories in a basin of attraction. One place where the study of focus is quite advanced concerns entrainment or synchronization (Strogatz has provided a very good popular survey of the field[24]). The tendency of systems to move to an entrainment pattern once past a threshold is described by Deleuze as a 'quasi-cause'.[25] The question is why this tendency should be granted an ontologically distinct status from the action of the negative feedback mechanisms. One of the key issues here is multiple realizability, the way patterns which are modelled by identical singularity layouts in manifolds can appear in widely different instanciations. Following Lautman's ontological distinction between singularities and the trajectories which are shaped by them, Deleuze says yes, here we need an ontological difference: the patterns defined by a layout of singularities in a manifold should be called virtual multiplicities, because they structure many spatio-temporally distinct intensive morphogenetic processes that result in widely different actual products. It's the quasi-causal action of these multiply realizable (in Deleuze's terms, differencially actualizable) patterns, patterns which channel behaviour and which are modelled in basins of attraction, which must be distinguished from the efficient, billiard ball, causality of negative feedback loops operating in intensive states of actual systems. This is the reformed notion of final causality to which DeLanda and Juarrero refer,[26] by means of which we can avoid the false problem of downward causation, when that is conceived as efficient causality emanating from a reified totality. But only when we study the onset of synchronic emergence does the notion of quasi-cause make sense; hence we need to think diachronic emergence.

Diachronic emergence, transversal emergence and political physiology

If one were to stay with the perspective of synchronic emergence, one would indeed find a hierarchy of material systems, so that individuals on one level are components of emergent unities on the next level: cell, organ, somatic body, social body.[27] The organism or subject is one level of this hierarchy, though we can go below the subject to a multiplicity of agents or independent behaviour patterns that when triggered are run on the system's organic hardware (to indulge in a little computer jargon that is actually out of place for technical reasons that have to do with the presuppositions of the computationalist versus embodied mind paradigms). In any event, for DG the multiplicity of agents as component parts of a subject is not simply equivalent to the 'society of mind' thesis, which deals only with the composition of a cognitive architecture of abstract functions.[28] Rather they would look at brain-level cognitive multiplicity by recourse to neurological findings and to the notion of the modularity of subsystems, or as they would call them in *Anti-Oedipus*, 'desiring machines'. An example of such a modular agent would be rage, a primitive mammalian inheritance.[29] In *A Thousand Plateaus* they describe it this way: 'schizoanalysis (. . .) treats the unconscious as an acentered system (. . .) as a machinic network of finite automata' (*ATP*, 18).

But the perspective of diachronic emergence shows that the time scales of each level are staggered, so that what appears as a systematic unity on a specific level is an event, a process, from the perspective of another level with a longer time scale. We can call this heterochrony: cells come and go but the organ stays (relatively) the same; people die but the social body lives on, and so on. Now we must also remember that 'above' the subject dimension there are intermediate scales: it is not just a question of 'individual and society', but of lots of not always harmoniously competing institutions, groups, bands, networks, nations, families, dysfunctional couples — you name it. DeLanda's work on social ontology in *A Thousand Years of Nonlinear History* is indispensable here: hierarchies and meshworks (strata and consistencies, to use DG's terms), but also meshworks of hierarchies and hierarchies of meshworks.[30]

Finally, we need to think how DG enable us to think emergence 'transversally' in their concept of assemblages. The key addition of DG is the focus on transversality, the thought of assemblages, which renders more complex the already complex notion of a heterochronous hierarchy sketched above. For example, the eukaryotic cell, which one might propose as the base level in a synchronic reading of the human organism, is itself already an organic assemblage according to the symbiogenesis theory of Lynn Margulis, who proposes, across evolutionary time scales, the incorporation of mitochondria — previously independent bacteria — into the emergent unity of the eukaryotic cell.[31] In *A Thousand Plateaus*, DG show how social machines intersect technological lineages in producing 'machinic phyla', or groups of assemblages defined by their affects: what they can do and what they can undergo. Thus the horse-man-stirrup assemblage of the steppe nomads also produces a bio-social-technical functional unit that is no simple aggregate. These assemblages are territorialized: the triggers of self-organizing behaviour are embedded systematically in a territory. The territorial assemblage interweaves a machinic assemblage of bodies and a collective assemblage of enunciation so that behaviour patterns are reliably triggered given the utterance of 'order words' (*ATP*, 88). A territorial assemblage produces emergent unities 'tranversally' among organisms, subjects and technological apparatuses.

As the above examples indicate, our three forms of emergence admit of some further nuances, as transverse emergence can be either homeostratic or heterostratic as well as synchronic or diachronic. To be precise, the isolation of an assemblage as containing only entities of the same stratum is an artificial selection: all organic entities have close ties to their inorganic milieus, while all social assemblages are technical and organic at the same time. Nonetheless, let us present them in outline form, with apologies for the rather barbaric terminology:

1. Homeostratic synchronic transversal emergence
   a. organic (symbiosis among organisms; ecosystems among groups of organisms)
   b. social (institutions forming a larger entity: e.g., colleges forming a university)
   c. technical (e.g., computers and routers forming the Internet)
2. Homeostratic diachronic transversal emergence
   a. organic (symbiogenesis: Margulis's theory of the origin of the eukaryotic cell)
   b. social (system change: e.g., change of the university from education of an elite into a centre for mass vocational training/military-industrial research)
   c. technical (system change: e.g., from ARPANET to Internet to world wide web)
3. Heterostratic synchronic transversal emergence (a bio-social-technical assemblage)
4. Heterostratic diachronic transversal emergence (mutation and co-evolution of such assemblages in 'machinic phyla')[32]

Moving us above, below, and diagonal to the organism or subject, DG thus enable us to construct a concept of 'political physiology' which studies the way interlocking intensive processes articulate the patterns, thresholds, and triggers of emergent bodies, forming assemblages linking the social and the somatic, with sometimes — but not always — the organic or subjective as intermediary. Sometimes we have direct links of the social and the sub-organismic or sub-subjective, as when, for example, American military training embeds 'shoot to kill' reflexes in the spinal cords of GIs, reflexes triggered by the presentation of human-shaped targets in the appropriate 'free fire zone'.[33]

The organism and emergence

Let us see how all these notions of emergence intersect the question of the organism as conceived by DG. As a fruitful contrast to DG, let us think of the work of Francisco Varela, both in his early work with Humberto Maturana ('autopoiesis') and later in his contribution to the 'embodied mind' school of cognitive science ('enaction').[34]

Briefly put, DG will completely agree with the autopoietic notion of the organism as an instance of synchronic emergence dedicated to homeostatic stability, but they want also to think the relation of the (actual) organism to life, which for them is a virtual multiplicity which is actualized in differenciating bursts of diachronic emergence — a notion which, it turns out, is quite close to the 'natural drift' argument of the later Varela. In addition, they also have a notion of 'non-organic life', which would be their way of talking not simply of inorganic self-organization (that is, homeostratic synchronic emergence below the organic level[35]), but also heterostratic transversal emergence. There are thus two critiques of the 'organism' we need to distinguish. The first simply points out the difference between having an identitarian or differential horizon for thinking change. The second is not so much a complaint about homeostasis in individual somatic bodies as it is a critique of what they call Oedipus, a particular way in which human organisms have come to be knit into the bio-social-technical assemblages of capitalist society. In other words, the second critique of the organism is also a call to think heterostratic transversal emergent assemblages, or political physiology.

In *A Thousand Plateaus*, the organism is a centralized, hierarchized, self-directed body, the 'judgment of God' (He who provides the model of such self-sufficiency), a molarized and stratified life form.[36] The organism is a synchronic emergent effect of organizing organs in a particular way, a 'One' added to the multiplicity of organs in a 'supplementary dimension' (*ATP*, 21, 265). The organism is the unifying emergent effect of interlocking homeostatic mechanisms that quickly compensate for any non-average fluctuations (below certain thresholds, of course) to return a body to its 'normal' condition (as measured by species-wide norms). As a stratum, we can use DG's specialized terminology for the organism. Skipping over several scales (cell, tissue, organ) for simplicity's sake, we arrive at the level of organic systems (e.g., the nervous, endocrine, and digestive systems), where the substance of content is composed of organs and the form of content is coding or regulation of flows within the body and between the body and the outside. The form of expression at this level is homeostatic regulation (overcoding of the regulation of flows provided by organs), while the substance of expression, the highest-level emergent unifying effect, is the organism, conceived as a process binding the functions of a body into a whole through co-ordination of multiple systems of homeostatic regulation.

DG's reading matches the autopoietic conception of the organism. Autopoietic theory distinguishes between the (virtual) organization and the (actual) structure of organisms. Organization is the set of all possible relationships of the autopoietic processes of an organism; it is hence equivalent to a virtual field or the Body without Organs of that organism.[37] Structure is that selection from the organizational set that is actually at work at any one moment.[38] Perturbation from the environment in 'structural coupling' leads to structural changes which either re-establish homeostasis or result in the destruction of the system qua living system.[39] Homeostatic restoration thus results in conservation of autopoietic organization.

The difficulty here is that the assumption of organization as a fixed transcendental identity horizon prevents us from thinking life as the virtual conditions for creative novelty or diachronic emergence. Life for autopoiesis is restricted to maintenance of homeostasis; creative evolutionary change is relegated to structural change under a horizon of conserved organization. If virtual organization is conserved for each organism, no matter what the changes in its actual structure are — one of the prime tenets of autopoietic theory — then on an evolutionary time scale, all life has the same organization, that is, belongs to the same class, and has only different structure. As Hayles put it: 'either organization is conserved and evolutionary change is effaced, or organization changes and autopoiesis is effaced'.[40] Autopoietic theory gladly admits all this: 'Reproduction and evolution do not enter into the characterization of the living organization.'[41] Evolution is the 'production of a historical network in which the unities successively produced embody an invariant organization in a changing structure'.[42]

Although autopoietic theory, developed in the 1970s at the height of the molecular revolution in biology, performed an admirable service in reasserting the need to think at the level of the organism, it is clear that autopoiesis is locked into a framework which posits an identity-horizon (organizational conservation) for (structural) change. The critique of identity-based thinking developed in *Difference and Repetition* posits life as virtually creative, that is, posits pure difference or differentiation/differenciation as the horizon for change. For autopoietic theory, living systems conserve their organization, which means their functioning always restores homeostasis; evolution is merely structural change against this identity horizon. For Deleuze, life is virtual differentiation ceaselessly differenciating in divergent actualization; the self-identity of the organism, preserved by homeostasis, is just an expression of the necessity of dipping into actuality in order to provoke the next burst of virtual creativity. (There is a parallel argument about species and organic form, developed in Ansell Pearson's *Germinal Life* and responded to by Hansen,[43] which we treat below.)

The concept of the organism as enactive in Varela's later work is much closer to DG's interests in describing diachronic transversal emergence, as here we loop into the environment.[44] Perception is no longer simply triggering of response, but comes from action in the environment, from movement. Cognitive structures emerge from recurrent sensory-motor patterns.[45] Cognition, perception, and moving action are all intertwined in 'structural coupling' with the world and are all needed for development. Each organism enacts a world for itself in bringing forth what is significant to it, what has 'surplus signification'.[46]

On the macro-evolutionary time-scale, natural drift, which encompasses self-organization, structural coupling and bricolage of modules,[47] is the analogue of cognition as embodied action.[48] In natural drift, diversity of organic form shapes and is shaped by coupling with environment; it is an 'emergent property'.[49] Here survival and reproduction are only global constraints, for natural selection only rejects what doesn't meet these criteria, instead of selecting for an optimal fit of organism and environment. Thus in the concept of natural drift there is still a lot of room for what Deleuze in *Difference and Repetition* would call differenciation. The evolutionary problem is not finding optimality but pruning away diversity to allow substrata for satisficing solutions rather than optimizing what fits a pregiven external standard.[50] The problem posed to natural drift in *The Embodied Mind*, 'how to prune the multiplicity of viable trajectories',[51] thus comes much closer to Deleuze's notion of differenciation as diachronic emergence than was ever possible with autopoietic theory.

However, things are not so simple when we concentrate more closely on questions of transversal emergence and organic form. In dialogue with Ansell Pearson, Hansen points to the need to distinguish time scales; he claims DG conflate the macro-evolution time scale of symbiogenesis with the time scale of ontogeny or individual development. At stake is the relation of the virtual field (*the* BwO) of life as self-ordering and creative to the portion of the virtual field relevant to an organism (*a* BwO, the BwO of the organism, as what this particular body 'can do'). Hansen claims that DG neglect the constraints of the viability of organic form in favour of what he calls their 'cosmic expressionism'. That is, Hansen believes DG postulate a completely open 'molecular' field for any possible combination of organic forms, without regard to the conservation of viable organic form in species-wide norms. Thus even organic form is a haecceity, or arbitrary selection from an open and heterogenous virtual field. The key complaint is that DG consider individuation as (synchronic) haecceity while neglecting diachronic emergence from a morphogenetic field, which needs the constraint of natural kinds channelling development.

Hansen is onto something here, which we can see when we consider the following lines from *A Thousand Plateaus*: 'The BwO howls: "They've made me an organism! They've wrongfully folded me! They've stolen my body!"' (*ATP*, 159). As a fixed habitual pattern locked onto normal functioning as determined by species-wide average values, the organism deadens the creativity of life, the possibilities of diachronic emergence; it is 'that which life sets against itself in order to limit itself' (503). The organism is a construction, a certain selection from the virtual multiplicity of what a body can be, and hence a constraint imposed on the BwO (the virtual realm of life, the set of all possible organic forms). Like all stratification, however, the organism has a certain value: 'staying stratified — organized, signified, subjected — is not the worst that can happen' (161), although this utility is primarily as a resting point for further experimentation, the search for conditions that will trigger diachronic emergence. Hansen is correct here: DG's insistence on caution in experimentation only recognizes individual organism survival as a negative condition for further experiment, without giving any positive role to the organism-level self-organizing properties of the morphogenetic field.

The only thing to say on behalf of DG is that Hansen's analysis is organic, all too organic: it is homeostratic. While the statement 'staying stratified is not the worst thing' is negative, we also have to remember that all the strata intermingle (*ATP*, 69), that the body is a body politic, that in the context of heterostratic transversal emergence, 'organism' is a political term. 'Organism' refers to body patterns being centralized so that 'useful labour is extracted from the BwO' (159). We see that 'organism' is a term of political physiology when we realize that for DG the opposite of the organism is not death, but 'depravity': 'You will be an organism (. . .) otherwise you're just depraved' (159). That is, being an organism means that your organs are Oedipally patterned for hetero-marriage and work. Getting outside the organism doesn't mean getting outside homeostasis guaranteed by a certain organic form so much as getting outside Oedipus into what Oedipal society calls 'depravity'. Furthermore, political physiology or the thought of the body politic means that we have to think *le corps* as social in the context of heterostratic transversal emergence: we are all a Corps of Engineers. When a body links with others in a bio-social-technical assemblage it is the complex, transversally emergent body that increases *its* virtual realm, that is, it is the BwO of the bio-social-technical body that is at stake, not that of each individual organism or somatic body. So the experimentation DG call for is not so much with somatic body limits (although that is part of it) but with bio-social-technical body relations in heterostratic diachronic transversally emergent assemblages, what DG will call a 'consistency' or a 'war machine'.

Conclusion: toward a discussion of the subject and emergence

Using this article as a basis, in another context I hope to discuss the way DG's notions of emergence can help us to see that a discussion of the genesis of agents, rational and otherwise, can help in many of the debates in social science on methodological individualism, sociological realism, Rational Choice Theory, the structure/agency dilemma, and other problematic areas. In doing so we will have to take account of the difference between the synchronic emergence of stable systems, with their equilibrium focus, and the ability to sustain creativity in what DG call consistencies or war machines: 'the nomad reterritorializes on deterritorialization itself' (*ATP*, 381). Such absolutely deterritorializing assemblages are not simply resilient and creative, but are so precisely in ways that maintain the conditions for future creativity; they find ways to stay in their crisis zones, they feel at home while on the move — even if they don't move relative to their position in a GPS system. In other words, their pattern allows them to change their pattern. We can thereby dissolve the false dichotomy between social holism (oriented to homeostatic stability) and methodological individualism (which denies ontological emergence), as well as evade the antinomies of the structure/agency debate, by showing that agency, when conceived as creativity in changing the patterns and thresholds of social systems, can only appear in far-from-equilibrium crisis situations.[52]

NOTES

1 In recent and noteworthy attempts, Richard McDonough provides a taxonomy of 27 (!) concepts of emergence ('Emergence and Creativity: Five Degrees of Freedom', in *Creativity, Cognition and Knowledge: An Interaction*, edited by Terry Dartnell (Westport CT, Praeger, 2002), 283–320).

Sami Pihlstrom provides an excellent overview of the field and a pragmatist defence of emergence ('What Shall We Do With Emergence? A Survey of a Fundamental Issue in the Metaphysics and Epistemology of Science', *South African Journal of Philosophy* 18:2 (May 1999), 192–211). Jürgen Schröder, on the other hand, will confine emergence to an epistemic category (Jürgen Schröder, 'Emergence: Non-Deducibility or Downwards Causation?', *The Philosophical Quarterly* 48:193 (October 1998), 433–52). See also David Spurrett, 'Bhaskar on Open and Closed Systems', *South African Journal of Philosophy* 19:3 (2000), 188–209, and R. Keith Sawyer, 'Emergence in Sociology: Contemporary Philosophy of Mind and Some Implications for Sociological Theory', *American Journal of Sociology* 107:3 (November 2001), 551–85.

2 Deleuze and Guattari's teamwork in producing a number of co-authored works is itself an example of emergence. I will call the emergent unity 'DG'. One of the things contemporary philosophy needs most is a discussion of the relations between Deleuze's work, Guattari's work, and the work of DG. See Gilles Deleuze and Félix Guattari, *Anti-Oedipus*, translated by Robert Hurley, Mark Seem and Helen R. Lane (London, Athlone, 1984); *Kafka: Toward a Minor Literature*, translated by Dana Polan (Minneapolis, University of Minnesota Press, 1986); *A Thousand Plateaus: Capitalism and Schizophrenia*, translated by Brian Massumi (Minneapolis, University of Minnesota Press, 1987) — henceforth ATP; *What is Philosophy?*, translated by Hugh Tomlinson and Graham Burchell (New York, Columbia University Press, 1994).

3 Manuel DeLanda, *Intensive Science and Virtual Philosophy* (London, Continuum Press, 2002); Gilles Deleuze, *Difference and Repetition*, translated by Paul Patton (New York, Columbia University Press, 1994); Gilles Deleuze, *The Logic of Sense*, translated by Mark Lester with Charles Stivale (New York, Columbia University Press, 1990). See also Brian Massumi, *A User's Guide to Capitalism and Schizophrenia: Deviations from Deleuze and Guattari* (Cambridge MA, MIT Press, 1992), though he focuses on DG. The question of Deleuze's materialism and/or realism — or in other words, the relation of philosophy and science in his thought (Queen, handmaiden, or something else entirely) — is extremely complex. As the better part of valour I will defer this issue to another context, though the confrontation with the position established by James Williams in *Gilles Deleuze's Difference and Repetition: A Critical Introduction and Guide* (Edinburgh, Edinburgh University Press, 2003) — namely that mathematics and science are for Deleuze mere illustrations of a fundamentally philosophic stance — will be essential to that endeavour.

4 Mark Bonta and I provide an overview of Deleuze and complexity theory, with extensive references, in our *Deleuze and Geophilosophy: A Guide and Glossary* (Edinburgh, Edinburgh University Press, 2004).

5 See Silberstein and McGeever and Thompson and Varela for a discussion of how thinking in terms of complex systems studied by means of non-linear dynamics changes the terms of the classic debate summarized by McDonough, which remains burdened with an impoverished view of matter (Michael Silberstein and John McGeever, 'The Search for Ontological Emergence', *The Philosophical Quarterly* 49:195 (April 1999), 182–200; Evan Thompson and Francisco Varela, 'Radical Embodiment', 418–25). See also McDonough, 'Emergence and Creativity'. Jeffrey Goldstein also provides a sympathetic account of emergence in terms of complexity theory ('Emergence as a Construct: History and Issues', *Emergence* 1:1 (1990), 49–72). I deal with hylomorphism extensively in John Protevi, *Political Physics: Deleuze, Derrida and the Body Politic* (London, Athlone Press, 2001).

6 In *Intensive Science and Virtual Philosophy* DeLanda attempts to outflank the entire reductionism question by proposing a Deleuzian epistemology that redefines science from the search for laws in nature to the search for topological regularities in scientific fields, or as he puts it, the distribution of singular and ordinary points in a problem (127–8).

7 Silberstein and McGeever, 'The Search for Ontological Emergence', 197.

8 DeLanda, *Intensive Science and Virtual Philosophy*, 80, 110, 126.

9 See Alicia Juarrero, *Dynamics in Action: Intentional Behavior as a Complex System* (Cambridge MA, MIT Press, 1999), 127, 143.

10 Here we adapt terms from autopoietic theory, which is discussed below.

11 Mark Bonta and I treat this theme at length in our *Deleuze and Geophilosophy*. See also Francisco Varela, 'Organism: A Meshwork of Selfless Selves', in *Organism and the Origins of Self*, edited by Alfred I. Tauber (The Hague, Kluwer, 1991), 96.

12 See DeLanda, *Intensive Science and Virtual Philosophy*, 31. Here, he points out the importance for Deleuze of the work of Albert Lautman concerning this point.

13 See Silberstein and McGeever, 'The Search for Ontological Emergence'. As they note, the most conservative commentators will insist that chaos is a property of the model and that 'one cannot in general prove that a real physical system is chaotic in the rigorous sense in which a mathematical model is chaotic' (195). This sort of extreme caution seems to be a holdover from the positivist allergy to ontological commitment; it would take us too far afield to deal further with this question of realism and anti-realism, but we will note the strong realism of DeLanda's *Intensive Science and Virtual Philosophy*.

14 David Waldner, 'Anti Anti-determinism: or What Happens when Schrodinger's Cat and Lorenz's Butterly meet Laplace's Demon in the Study of Political and Economic Devlopment' (paper presented at the annual meeting of the *American Political Science Association*, August 29–September 1,

2002. Available online at: http://www.people.virginia.edu/~daw4h/Anti-Anti-Determinism.pdf).

15 Evan Thompson, Francisco Varela, 'Radical Embodiment: Neuronal Dynamics and Consciousness', *Trends in Cognitive Science* 5 (2001), 418–25.

16 The same phrase appears as the title of Ilya Prigogine and Isabelle Stengers's *Order Out of Chaos* (New York, Bantam Books, 1984), and the subtitle of Steven Strogatz's *Sync: How Order Emerges From Chaos in the Universe, Nature, and Everyday Life* (New York, Hyperion, 2003).

17 In 'Emergence and Creativity' McDonough usefully engages with Robert Klee ('Micro-Determinism and Concepts of Emergence', *Philosophy of Science*, 51:1 (March 1984), 44–63) as a classic case of anti-emergentism in the philosophy of mind.

18 For a critique of genetic determinism from an analytic philosopher, see Sahotra Sarkar, *Genetics and Reductionism* (Cambridge, Cambridge University Press, 1998). For the anti-genetic determinism of the Developmental Systems Theory (DST) school, see Paul Griffiths and R. Gray, 'Developmental Systems and Evolutionary Explanation', *Journal of Philosophy* 91:6 (1994), 277–305.

19 Immanuel Kant, *Critique of Judgement*, translated by J.H. Bernard (New York, Hafner, 1972).

20 See Schröder, 'Emergence: Non-Deducibility', and Sawyer, 'Emergence in Sociology', 560.

21 Sawyer, 'Emergence in Sociology', 560.

22 Silberstein and McGeever, 'The Search for Ontological Emergence', *The Philosophical Quarterly* 49:195 (April 1999), 182–200.

23 Sawyer, 'Emergence in Sociology'.

24 Strogatz, *Sync*.

25 See DeLanda, *Intensive Science and Virtual Philosophy*, 78–80.

26 *Intensive Science and Virtual Philosophy*, 80, 110, 126; Juarrero, *Dynamics in Action*, 127, 143.

27 By 'social body' I mean that human groups — institutions, teams, corporations, families, and so on — can be seen as emergent unities, with a focused systematic behaviour whose powers extend beyond that expected by simply adding up the power of members acting alone. This is of course the thesis denied by methodological individualism in social science, which is treated in Bonta and Protevi's *Geophilosophy*.

28 Francisco Varela, Evan Thompson and Elizabeth Rosch, *The Embodied Mind* (Cambridge MA, MIT Press, 1991).

29 See Paul Griffiths, *What Emotions Really Are: The Problem of Psychological Categories* (Chicago, University of Chicago Press, 1997). Griffiths calls rage — along with other primitive emotions — an 'affect program'.

30 Manuel DeLanda, *A Thousand Years of Nonlinear History* (New York, Zone Books, 1997).

31 Lynn Margulis and Dorion Sagan, *Microcosmos: Four Billion Years of Microbial Evolution* (Berkeley, University of California Press, 1986).

32 On bio-social-technical assemblages or 'cyborgs' in the context of cognitive science and philosophy of mind, see Andy Clark, *Natural Born Cyborgs* (Oxford, Oxford University Press, 2003). On the notion of machinic phyla, see Deleuze and Guattari, *A Thousand Plateaus*, 404–10.

33 See David Grossman, *On Killing* (Boston, Little, Brown, 1996). Grossman has details of such reflex training, which was first implemented in the Vietnam era to increase the firing rates of infantrymen, which were found to be unacceptably low in post-WWII studies. I have treated the issue at length in 'Political Physiology in High School: On Columbine', in *Deleuze and the Contemporary World*, edited by Adrian Parr and Ian Buchanan (Edinburgh, Edinburgh University Press, 2006).

34 See Humberto Maturana and Francisco Varela, *Autopoiesis and Cognition*, Boston Studies in the Philosophy of Science, vol. 42 (Boston, Reidel, 1980). This is the classic statement of autopoietic theory. For the embodied mind, see Varela, Thompson and Rosch, *The Embodied Mind*. For an overview of Varela's career, see David Rudrauf, Antoine Lutz, Diego Cosmelli, Jean-Philippe Lachaux and Michel Le Van Quyen, 'From Autopoiesis to Neurophenomenology: Francisco Varela's Exploration of the Biophysics of Being', *Biological Research* 36:1 (2003), 27–65.

35 As in the self-organization of lipids to form membranes, a common concept in theories of the origin of life.

36 I detail DG's notion of the organism in relation to Aristotle and Kant in 'The Organism as the Judgment of God: Aristotle, Kant and Deleuze on Nature (that is, on Biology, Theology and Politics)', in *Deleuze and Religion*, edited by Mary Bryden (London, Routledge, 2001), 30–41.

37 Maturana and Varela, *Autopoieses and Cognition* mentions autopoietic 'space' (scare quotes in the original).

38 See Maturana and Varela, *Autopoiesis and Cognition*, xx, 77, 137–8. See also N. Katherine Hayles, *How We Became Posthuman* (Chicago, University of Chicago Press, 1999), 138; Rudrauf et al., 'From Autopoiesis to Neurophenomenology', 31.

39 Maturana and Varela, *Autopoiesis and Cognition*, 81.

40 Hayles, *How We Became Posthuman*, 152.

41 Maturana and Varela, *Autopoiesis and Cognition*, 96.

42 *Autopoiesis and Cognition*, 104.

43 Keith Ansell Pearson, *Germinal Life: The Difference and Repetition of Deleuze* (London, Routledge, 1999); Mark Hansen, 'Becoming as Creative Involution?: Contextualizing Deleuze and Guattari's Biophilosophy', *Postmodern Culture* 11:1 (September 2000), http://www3.iath.virginia.edu/pmc/text-only/issue.900/11.1hansen.txt.

44 See Varela, 'Organism: A Meshwork of Selfless Selves' and Varela, Thompson and Rosch, *The Embodied Mind*.

45 Varela, Thompson and Rosch, *The Embodied Mind*, 173–9; see also George Lakoff and Mark Johnson, *Philosophy in the Flesh* (New York, Basic Books, 1999).

46 Varela, 'Organism: A Meshwork of Selfless Selves', 96.

47 Varela, Thompson and Rosch, *The Embodied Mind*, 197.

48 *The Embodied Mind*, 188.

49 *The Embodied Mind*, 195.

50 *The Embodied Mind*, 196.

51 *The Embodied Mind*, 196.

52 Of course it is never easy to identify the criteria by which one can recognize a crisis. See David Waldner's paper 'Anti Anti-determinism: or What Happens when Schrodinger's Cat and Lorenz's Butterly meet Laplace's Demon in the Study of Political and Economic Development' (presented at the annual meeting of the *American Political Science Association*, August 29–September 1, 2002. Available online at: http://www.people.virginia.edu/~daw4h/Anti-Anti-Determinism.pdf). This paper offers a treatment of the political science literature on 'regime change' with regards to this issue.

Chaosmologies: Quantum Field Theory, Chaos and Thought in Deleuze and Guattari's *What is Philosophy*?

ARKADY PLOTNITSKY

Abstract
This article explores the relationships between the philosophical foundations of quantum field theory, the currently dominant form of quantum physics, and Deleuze's concept of the virtual, most especially in relation to the idea of chaos found in Deleuze and Guattari's *What is Philosophy?*. Deleuze and Guattari appear to derive this idea partly from the philosophical conceptuality of quantum field theory, in particular the concept of virtual particle formation. The article then goes on to discuss, from this perspective, the relationships between philosophy and science, and between their respective ways of confronting chaos, a great enemy but also a great friend of thought, and its greatest ally in its struggle against opinion.

Keywords: chaos, concept, philosophy, quantum field theory, science, thought, virtual

Chaos and concepts in physics and philosophy

The aim of this essay is to explore the relationships between the philosophical underpinnings of quantum field theory and Gilles Deleuze's concept of the virtual, specifically in conjunction with the idea of chaos found in Gilles Deleuze and Félix Guattari's *What is Philosophy?* The book approaches chaos by means of a particular and, in philosophy, rarely, if ever, used concept. According to Deleuze and Guattari: 'Chaos is defined not so much by its disorder as by the infinite speed with which every form taking shape in it vanishes. It is a void that is not a nothingness but a *virtual*, containing all possible *particles* and drawing out all possible forms, which spring up only to disappear immediately, without consistency or reference, without consequence. Chaos is an infinite speed of birth and disappearance' (emphasis on 'particles' added).[1] Deleuze and Guattari will go on to argue that art, science, and philosophy are different, if sometimes interrelated, ways in which thought confronts chaos — a great *enemy*, but also a *friend* of thought and its most important ally in its yet greater struggle, that against opinion, *doxa*.[2]

Although unusual elsewhere, this type of idea of chaos, or at least of *virtuality*, appears in part to be derived by Deleuze and Guattari from quantum field theory, as is also suggested by their reference to Ilya Prigogine and Isabelle Stengers' *Entre le temps et l'éternité*,[3] from which they borrow the expression 'all possible particles' (*WP*, 225, note 1). The idea relates to the so-called virtual birth and disappearance, or creation and annihilation, of particles from the so-called false vacuum, a kind of sea of energy, thus suggesting the image of chaos invoked by Deleuze and Guattari, although the term chaos itself is not used in quantum field theory in this context.

These connections between Deleuze and Guattari's concept of chaos and quantum field theory also allow me to link the idea of *chaos as the virtual* to the idea of *chaos as the incomprehensible*, which can be traced to the Ancient Greek idea of chaos as *areton* or *alogon* — that which is beyond all comprehension. This link arises from the possibility that the processes responsible for the creation or annihilation of forms, for their birth and disappearance, or for the speed of both, may not be representable or even conceivable by any means available to us. Deleuze and Guattari's invocation of black holes (whose ultimate constitution is beyond our comprehension) and surrounding elaborations in *A Thousand Plateaus* supports the significance of this concept of chaos (coupled to chaos as the virtual) in their work.[4] In addition, the connections to quantum field theory allow me to bring into consideration yet another concept of chaos, *chaos as disorder*, defined by the role of chance in its workings or in its effects. As Deleuze and Guattari's formulation indicates, the concept of chaos as disorder is not entirely put aside by them: while 'chaos' may be 'defined *not so much* by its disorder', it may partially be defined by disorder or, at least, by chance. Instead, this concept of chaos as chance and disorder, or the concept of chaos as the incomprehensible, is to some degree subordinated to the concept of chaos as the virtual. The same is true in quantum field theory, which takes over the conceptions of chaos as the incomprehensible and chaos as chance and disorder from quantum mechanics, but adds to them the concept of chaos as the virtual and gives it an analytically dominant role.[5]

One could also see at least some of the *concepts* of quantum field theory as *philosophy*, in accordance with or close to Deleuze and Guattari's argument in *What is Philosophy?* According to Frank Wilczek, a leading contemporary quantum-field theorist and a Nobel Prize laureate, 'the primary goal of fundamental physics is to discover profound concepts that illuminate our understanding of nature'.[6] The concepts Wilczek has primarily in mind are physical concepts, and they must be, given the disciplinary character of modern physics as a mathematical-experimental science. These concepts may, however, also be seen as concepts in terms of Deleuze and Guattari's concept of concept, introduced in the book. A philosophical concept in this sense is not an entity established by a generalization from particulars or 'any general or abstract idea' (*WP*, 11–12). Instead it is an irreducibly complex, multi-layered structure or architecture — a multi-component conglomerate of concepts in their conventional sense, figures, metaphors, particular elements, and so forth — and as such may define a whole philosophical matrix. As they say, 'there are not simple concepts' (15). Philosophy itself is defined by Deleuze and Guattari as a creation of new concepts and even concepts that are 'always new', thus making it, in Nietzsche's phrase, the philosophy of the future (5).

Deleuze and Guattari's description of chaos and accompanying elaborations on science (to be considered below) have been a subject of some controversy during the so-called 'Science Wars', especially in the wake of Alan Sokal and Jean Bricmont's *Impostures intellectuelles*, originally published in 1997, which considers some of these elaborations.[7] Since I have discussed the subject at length on a previous occasion, to which I permit myself to refer here, I shall restrict myself to a few essential points especially germane to the context of this article.[8] Sokal and Bricmont fail to offer an adequate reading of Deleuze and Guattari (or other authors they discuss) primarily because they miss or bypass the architecture of their philosophical concepts, defined, as I explained, by complex mixtures or *mélanges*, including when science is used. They also miss the difference between science and philosophy, or their respective ways of dealing with chaos, which is, ironically, at stake in Deleuze and Guattari's book, including in the elaborations that Sokal and Bricmont cite, but do not really read. The more nuanced complexity of the interrelationships or 'interferences' between philosophy and science, including the philosophical dimensions of scientific concepts, would require a kind of reading of the overall argument of the book that Sokal and Bricmont appear to be unwilling to undertake. Their 'readings' usually amount to citing long passages and declaring them, at best, *mélanges* of sense and nonsense, while such passages require extensive exegeses, even if one wants to be critical, and especially if one does. I am not saying that one cannot criticize Deleuze and Guattari. I am saying, however, that Sokal and Bricmont do not appear or fail to prove themselves to be in a position adequately to discriminate between what is and is not an appropriate use of science in the texts they consider.

Thus, in commenting on a long passage from *What is Philosophy?* (119–20), they say: 'With a bit of work, once can detect in this paragraph a few meaningful phrases, but the discourse is which they are immersed is utterly meaningless' (*FN*, 158). '*Utterly* meaningless'! But they do not explain why. A footnote is added: 'For example, the statement "the speed of light (. . .) where length contracts to zero and clocks stop" is not false, but may lead to a confusion. In order to understand it correctly, one must already have a good knowledge of relativity theory' (158, note 202). That may be true, but it can hardly be held against Deleuze and Guattari, who report a correct scientific finding and report it correctly, as Sokal and Bricmont acknowledge. Deleuze and Guattari's actual point in the passage is that this peculiar physical situation is strictly linked to a specific number, the speed of light, 299,796 kilometers per second. The relationships between scientific concepts and measurable numerical quantities define modern science, as Deleuze and Guattari make clear by noting that, 'the entire theory of functions [which defines the practice science] depends on numbers' (*WP*, 119). This is hardly meaningless, let alone 'utterly meaningless', but it requires a reading of what Deleuze and Guattari actually say, which is not something Sokal and Bricmont ever offer. Under these circumstances, an intellectually and ethically appropriate claim on Sokal and Bricmont's part could have been that *they cannot make sense* of this or other passages in question but not that *these passages themselves make no sense*, as they contend.

Sokal and Bricmont's statements on science sometimes have problems of their own. Their commentaries on quantum theory (their field of expertise) are not always rigorous and helpful, and sometimes are technically inaccurate. Thus, the reader of their book may indeed be confused by their discussion of Schrödinger's equation, the fundamental equation of quantum mechanics, and its linearity. They never explain an important difference between linear equations in elementary algebra (which have a form $ax + b = c$) and linear *differential* equations, such as Schrödinger's equation, which a crucial point, since it affects what kind of process the latter accounts for (*FN*, 143–5). They also appear not to realize that imaginary and complex numbers are in fact irrational, while Jacques Lacan, whom they criticize on the subject, understands this irrationality much better (*FN*, 25).[9]

These problems have a positive role to play in reminding us that scientists are not always sufficiently accurate and do not always have sufficient expertise even in their own fields, and that we should not necessarily trust them on science. So much in Sokal and Bricmont's book, and by so many, was accepted merely on the strength of their authority as scientists and their declarations concerning science and its uses and abuses, declarations unsupported by arguments. Science itself and those nonscientific readers who want to learn about it are not served well by their book. Both are served much better by an engagement with the complexity of the relationships between philosophical and scientific thought, which *What is Philosophy?* pursues as part of its approach to the question its title asks and in moving beyond it.

Physics' chaosmologies

Before addressing quantum field theory itself, I shall, by way of a background, discuss, first, classical physics and then, quantum mechanics. Classical physics is defined by the fact that it considers its objects and their behaviour as available to conceptualization and to representation in terms of particular properties of these objects and their behaviour. Such properties are abstracted and these objects idealized from actual objects in nature for the purposes of connecting the mathematics of the theory to the measurable quantities constituting the experimental data. In Deleuze and Guattari's terms, this is the way in which classical physics creates its frame of reference and, thus, *slows down* the chaos of nature and its emerging and disappearing forms, encountered by our thought (*WP*, 118).

Classical or Newtonian mechanics (which deals with the motion of individual physical objects or systems composed of such objects) accounts for its objects and their behaviour on the basis of physical concepts, such as 'position' and 'momentum', and measurable quantities corresponding to them. Classical mechanics is, thus, ontologically, *realist* because it can be seen as fully describing all of the (independent) physical properties of its objects necessary to explain their behaviour. It is, ontologically, *causal* because the state of the systems it considers at any given point is assumed to be determined (in the past) by and to determine (in the future) its states at all other points. It is, epistemologically, *deterministic* insofar as our knowledge of the state of a classical system at any point is, first of all, possible and, secondly, allows us to know, again, at least in principle and in ideal cases its state at any other point.

Causal theories need not be deterministic. While conforming to the realist and causal model of classical mechanics at the ultimate level, other areas of classical physics, such as thermodynamics and statistical physics or chaos theory complicate the situation by introducing chance and, hence, a degree of *chaos as chance* into the picture. These theories are not deterministic, even in ideal cases, in view of the great structural complexity of the systems they consider. This complexity blocks our ability to predict the behaviour of such systems, either exactly or at all, even though we can write equations that describe them and assume their behaviour to be causal. Chaos theory is also realist, insofar as it maps the behaviour of material bodies, although deterministic predictions are not possible due to the complexity of the behaviour of the systems it considers. By contrast, classical statistical physics is not realist insofar as its equations, while allowing us to make correct statistical predictions, do not describe the behaviour of its objects, such as molecules of a gas. It is, however, based on the realist assumption of an underlying non-statistical multiplicity, whose individual members conform to the causal laws of Newtonian mechanics.

In order to conceptualize the Newtonian universe, Pierre-Simone de Laplace introduced his figure of a 'demon' — an intelligent being that controls the immense machinery of the universe (*WP*, 129). A more Newtonian 'conceptual persona' is that of God as a universal clock maker. Laplace's demon may be a better figure, since it brings chance into the picture at the human level. James Clerk Maxwell introduced his, equally famous, 'Maxwell's demon' in order to explain and connect chance and the underlying causality in his kinetic theory of gases. The introduction of chance into physics was a momentous event, not least because it introduced new ways of confronting and dealing with the chaos of our interactions with nature through thought, even if not of nature itself in its ultimate constitution, which remains causal prior to quantum theory.

Quantum mechanics, especially in certain interpretations (such as that of Niels Bohr, known as complementarity, which I follow here), is neither causal, nor deterministic, nor realist in any of the senses described above, in particular insofar as it makes it impossible to assign any specific form of independent physical reality to quantum objects or processes. By the same token, the theory is *fundamentally* statistical: it involves chance irreducible to any underlying causality. It is not only that the state of the system at a given point gives us no help in predicting its behaviour or allows us to assume it to be causally determined, if unpredictable, at later points, although that such is the case experimentally is important. More radically, this state itself cannot, at any point, be unambiguously defined on the model of classical physics: hence a lack of realism or the irreducible presence of chaos as the incomprehensible at the ultimate level. A lack of causality is an immediate consequence: for, if certain processes allow for no description of any kind, they would automatically disallow a causal description.

This impossibility of an unambiguous definition of the state of the system is correlative to Heisenberg's uncertainty relations. Technically, the latter express the (strict) quantitative limits, absent in classical physics, on the simultaneous joint measurement of the so-called—by analogy with classical physics—conjugate variables, which define the motion of classical objects, such as 'position' or 'coordinate' (q) and 'momentum' (p). These limits are given by the famous formula $\Delta q \Delta p \cong h$, where h is Planck's constant, and Δ designates the degree of imprecision of measurement. The increase in precision in measuring one such variable inevitably implies equally diminished precision in measuring the other. Bohr's complementarity gives a more radical interpretation to the uncertainty relations. This interpretation prohibits even an assignment or unambiguous definition of physical properties, such as a position or a momentum, to quantum objects and behaviour, rather than only establishing the limit (defined by Planck's constant, h) upon the degree of precision with which both can be simultaneously measured. Ultimately, such an assignment is impossible even if each such variable is taken by itself. All actual physical properties considered belong to the measuring instruments involved under the impact of quantum objects, and Heisenberg's formula now applies to these properties.

Deleuze and Guattari offer the following comment on Heisenberg's uncertainty relations, made in the context of demonic 'partial observers' in science, which 'Heisenberg's demon' of uncertainty symbolizes:

> To understand the nature of these partial observers that swarm through all the sciences and systems of reference, we must avoid giving them the role of a limit of knowledge or of an enunciative subjectivity. (...) As a general rule, the observer is neither inadequate nor subjective: even in quantum physics, Heisenberg's demon does not express the impossibility of measuring both the speed [more accurately, momentum] and the position of a particle on the grounds

> of a subjective interference of the measure with the measured, but it measures exactly an objective state of affairs. (*WP*, 129)

Bohr, too, stresses the objective character of all quantum mechanical observation and phenomena, and one can only speak of a 'limit of knowledge', in the sense that no knowledge of the kind classical mechanics offers as concerns its objects is available. Otherwise, quantum mechanics, *within its proper scope*, gives us as much knowledge as nature allows for. (Quantum field theory gives us more knowledge by expanding the scope of quantum theory.)

As indicated already, given the irreducible presence of chaos as the incomprehensible in quantum mechanics, its statistical character becomes objectively irreducible as well. The workings of the quantum-mechanical chance are fundamentally different from the classical picture of chance outlined above. The chance one encounters in quantum physics is irreducible not only in practice but also in principle. There is no knowledge in principle available to us, now or ever, that would allow us to eliminate chance and replace it with the picture of necessity behind it. Nor, however, can one postulate a causal dynamics as unknown or even unknowable but existing, in and by itself, outside our engagement with it. This qualification is crucial, since some forms of the classical understanding of chance allow for this type of (realist) assumption. 'Heisenberg's demon' makes chance and, hence, a certain element of disorder and of chaos as chance and disorder, unavoidable even in the case of individual elementary events (as against the classical view which relates chance to multiple processes and events), and in particular at this level. It can be shown that at the level of collective events quantum mechanics may exhibit more order than classical statistical physics does. What makes this chance or disorder irreducible is chaos as the incomprehensible, which links both conceptions of chaos in quantum mechanics.

Deleuze and Guattari's sense of chaos as the virtual still applies (as it does in classical physics), insofar as one still needs to find a way to handle the situation, physically and philosophically. Quantum mechanics predicts, and predicts *exactly*, the *probabilities* of the outcome of the relevant experiments, without (unlike classical physics) telling us anything about what happens in between or how such outcomes come about. In other words, chaos as the virtual only pertains to scientific *thought*, while chaos as the incomprehensible and chaos as chance also pertain to nature in the (chaosmic) order of quantum mechanics. Thus, all three forms of chaos — disorder, the incomprehensible, and (only at the level of thought) chaos as the virtual — are found, and are connected in quantum mechanics. Quantum field theory is different insofar as it adds chaos as the virtual to its chaosmology of nature itself, which was a momentous step in the history of physics. I shall now explain why it was compelled to move in this direction.

Suppose that one arranges for an emission of an electron from a source and then performs a measurement at a certain distance from that source. Merely placing a photographic plate at this point would do, and the corresponding traces could, then, be properly treated by means of quantum field theory. First, however, let us assume that we are dealing with the electron as a classical physical object. According to classical physics, one would encounter at this point the same object, and its position could be predicted exactly by classical mechanics. In quantum mechanics, by contrast, one would encounter either an electron or nothing, and quantum mechanics predicts the alternative probabilities for such events, for example, at fifty percent for each. This is why, as explained above, chaos as chance is unavoidable in quantum mechanics. Once the situation involves higher energies and is governed by quantum electrodynamics, the original form of quantum field theory, one may find an electron, nothing, a positron (anti-electron), a photon, an electron-positron pair, or, once we move to still higher energies or different domains governed by quantum field theory, still something else. The probabilities and only probabilities for the alternatives are properly predicted by quantum field theory, which makes chance and chaos as chance unavoidable in quantum field theory as well. The upshot is that in quantum field theory, an investigation of a particular type of quantum object (say, electrons) not only irreducibly involves other particles of the same type but also other types of particles, conceivably all existing types of particles. It is as if instead of an identifiable moving object of the type studied in classical physics, we encounter a continuous emergence and disappearance, creation and annihilation, of particles from point to point, the so-called *virtual* particle formation. While such events are in principle possible and their possibility defines the situation and what can and cannot *actually* occur, only some of them can be registered. Usually, those particles that are registered by observations are considered as 'real particles', while those that are not are considered as 'virtual particles'.

The corresponding quantum-field-theoretical *physical* concept possesses a mathematical and experimental rigour specific to science, while, however, retaining the key *philosophical* conceptual architecture of the virtual. Quantum field theory rigorously predicts which among such events can or cannot occur, and with what probability. All possible events are usually represented in terms of the so-called Feynman diagrams, after Richard Feynman, whose work brought him a Nobel Prize.[10] For example, the following diagram represents the annihilation and then the creation of an electron and a positron via a virtual photon (represented by a wavy line), with another virtual photon emitted by an electron later.

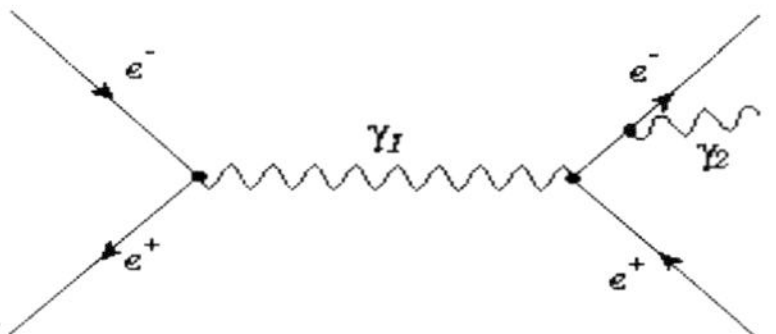

At any point of this diagram yet another virtual process (similar to the emission of a virtual photon γ_2 depicted here) may occur and hence another diagram may be inserted into it, thus leading to an interminably expandable rhizomatic structure of the type often invoked in Deleuze and Guattari's works. Every possible — virtual — event or transition can be represented by a Feynman diagram, and much of quantum field theory consists of drawing and studying such diagrams and generating predictions by using them.

Feynman diagrams are, however, just diagrams: they are pictures that help us heuristically to visualize the situation, or, in Deleuze and Guattari's idiom, to *slow down* the phenomenological chaos of the situation, to hold in mind the forms thus created, for the purposes of helping calculations. So is the 'picture' (conception) of the virtual particle formation. What actually happens at the level of such processes themselves we might no more know or even conceive of, let alone visualize, than we can in quantum mechanics, which implies the essential presence of chaos as the incomprehensible in quantum field theory. Since, in addition, all our knowledge concerning the ultimate constitution of nature is only predictive and, moreover, only statistical, chaos as chance and disorder is added to the picture as well. This is how chaos (of both quantum nature and of the mind confronting it), chaos as the incomprehensible and chaos as chance and disorder, was approached already by quantum mechanics, and it helped quantum mechanics against the *doxa* of classical physics. As it has often done before, physics had to plunge into the chaos to be able to create the order of quantum mechanics. Thus, ironically, *chaos*, chaos as the incomprehensible and chaos as chance or disorder, becomes part of a new *order* of quantum mechanics, or its chaosmos, to use James Joyce's coinage, favoured by Deleuze and Guattari. In quantum field theory, these two concepts of chaos are retained, but are not sufficient to deal with the chaos quantum field theory has to confront and to build its physical and mathematical architecture. To accomplish this task, it is compelled to engage with chaos as the virtual, to become a chaosmology of the virtual.

The concept of the virtual emerging from Deleuze's earlier works, such as *Difference and Repetition* and *The Logic of Sense*, onwards defines the virtual as *something that defines the space of what is possible and as such shapes the possible forms of the actual.*[11] This formulation is general and allows for different interpretations of the virtual, for example, on the model of classical physics, where the space of such possibilities is defined causally as the so-called phase space, which implies a realist and causal ontology, described above, and hence no chaos. This type of interpretation is found, for example, in Manuel DeLanda's work.[12] The Deleuzian virtual can also be interpreted on the model of quantum mechanics, which would prohibit a realist and causal ontology, and will introduce chaos as the incomprehensible and chaos as chance and disorder, but would not involve the concept of the virtual particle formation and, hence, no chaos as the virtual. This type of interpretation was advanced by Gilles Chatelet. It is, however, a quantum-field-theoretical concept of the virtual as chaos that, I argue, shapes the argument of *What is Philosophy?*[13]

'Casting planes over chaos': philosophy and science

Deleuze and Guattari define the difference between philosophy and science as follows:

> [P]hilosophy wants to know how to retain the infinite speed while gaining consistency, by *giving the virtual a consistency specific to it*. The philosophical sieve, as a plane of immanence that cuts through the chaos, selects infinite movements of thought and is filled with concepts formed like consistent *particles* going as fast as thought. Science approaches chaos in a completely different, almost opposite way: it relinquishes the infinite, infinite speed, in order to gain *a reference able to actualize the virtual*. By retaining the infinite, philosophy gives consistency to the virtual through *concepts*; by relinquishing the infinite, science gives a reference to the virtual, which actualizes it through *functions*. Philosophy proceeds with a plane of immanence and consistency; science with a plane of reference. (*WP*, 118; emphasis on 'particles', 'concepts' and 'functions' added)

The concept of the philosophical sieve is thus itself quantum-field-theoretical: concepts emerge in a plane of immanence similarly to the way virtual or actual particles are formed in and emerge from the false vacuum, a virtual energy field. Chaos as the virtual is itself such a philosophical concept. In its confrontation with chaos as the virtual, philosophy's thought may sometimes *hold* to a virtual 'particle' concept or (since concepts are not atomic, but have complex architectures) 'particle-conglomerate' concept, which can, if slowed-down or freeze-framed, be compared to a complex Feynman diagram. Deleuze and Guattari draw and work with such 'diagrams' of concepts, such as that of Descartes's concept of Cogito or a philosophical portrait of Kant (*WP*, 25, 56). In other words, at such junctures philosophy adopts science's way of dealing with chaos. But philosophy's thought would now hold to a concept that traverses a plane of immanence and, thus, gives this plane consistency. This is a complex philosophical concept, which cannot be further elaborated upon here. The main point is that, in contrast to philosophy, science 'freezes' chaos, or what it can in it, in slow motion or freeze-frames, such as Feynman diagrams in quantum field theory. Science, however, sometimes proceeds with the infinite speed of philosophy and, as just explained, philosophy sometimes proceeds by slowing the infinite speed of chaos down in the manner of science.

The difference between the two respective 'attitudes toward chaos', scientific and philosophical, is enabled, in the first place, by the difference in the determination of each in terms of functions and concepts, respectively.[14] Deleuze and Guattari write:

> The object of science is not concepts but rather functions that are presented as propositions in discursive systems. The elements of functions are called *functives*. A scientific notion is defined not by concepts but by functions or propositions. This is a very complex idea with many aspects, as can be seen already from the use to which it is put by mathematics and biology respectively. Nevertheless, it is this idea of the function that enables the sciences to reflect and communicate. Science does not need philosophy for these tasks. On the other hand, when an object — a geometrical space, for example — is scientifically constructed by functions, its philosophical concept, which is by no means given in the function, must still be discovered. Furthermore, a concept may take as its components the functives of any possible function without thereby having the least scientific value, but with the aim marking the differences in kind between concepts and functions. (*WP*, 117; translation slightly modified)

A philosophical concept corresponding to a mathematical or scientific object could be discovered by mathematics and science. Thus, it is a complex question of where and how, between mathematics (geometry and topology, which are already different), physics and philosophy, a modern, post-Riemannian or post-Einsteinian, philosophical concept of space has emerged. This concept can of course take on new aspects in philosophy qua philosophy, as it does in Deleuze's and Deleuze-Guattari's work. We are dealing here with a heterogeneous yet interactive *space* of relationships, where differences, similarities, and interactions are all found, but each becomes more or less crucial at different conceptual, historical, or cultural junctures. It is a dynamic space-time or a sea of energy of thought, a space at the edge of chaos — chaos as the incomprehensible, chaos as the virtual, and chaos as chance and disorder.

Deleuze and Guattari develop their argument concerning the difference between science and philosophy via the use of proper names in both fields. They write: 'If there is a difference between science and philosophy that is impossible to overcome, it is because proper names mark in one case a juxtaposition of reference and in the other [conceptual personae of philosophy] a superposition of layer: they are opposed to each other through all the characteristics of reference and consistency' (128). This argument forms a bridge to a 'demonology' of science, as part of its 'chaosmology', via the concept of the partial observer, mentioned earlier, which defines all scientific observation, although it is not to be identified with a human being and especially some subjective observer. They write:

We are referred back to another aspect of enunciation that applies no longer to proper names of scientists or philosophers but to their ideal intercessors internal to the domains under consideration. We saw earlier the philosophical role of *conceptual personae* in relation to fragmentary concepts on a plane of immanence, but now science brings to light *partial observers* in relation to functions within systems of reference. The fact that there is no total observer that, like Laplace's 'demon', is able to calculate the future and the past starting from a given state of affairs means only that God is no more a scientific observer than he is a philosophical [conceptual] persona. But 'demon' is still excellent as a name for indicating, in philosophy as well as in science, not something that exceeds our possibilities but a common kind of these necessary intercessors as respective 'subjects' of enunciation: the philosophical friend, the rival, the idiot, the overman are no less demons than Maxwell's demon or than Einstein's or Heisenberg's observers. It is not a question of what they can or cannot do but of the way in which they are perfectly positive, from the point of view of concept or function, even in what they do not know and cannot do. In both cases there is immense variety, but not to the extent of forgetting the different natures of the two great types. (128–9)

Although Deleuze and Guattari do not invoke the demons of quantum-field-theory, these demons hover over the book. The ways of dealing with chaos may be different in philosophy, art, and science. The concept of chaos is itself quantum-field-theoretical. The conclusion of *What is Philosophy?* envisions a possibility of a different future of thought, in which the boundary between philosophy, art, and science and even all three themselves disappears back into the chaosmic field of thought. The quantum-field-theoretical conception of chaos as the incomprehensible, as chance and disorder, and as the virtual remains in place and governs this vision as well.

'The shadow of "people to come"'

It is, then, their connections to chaos, their greatest enemy and their greatest friend, that make art, science, and philosophy so crucial to thought—against opinion, always an enemy only, 'like a sort of "umbrella" that protects us from chaos'. 'But', Deleuze and Guattari say, 'art, science, and philosophy require more: they cast planes over chaos' (*WP*, 202). What is more, 'the struggle with chaos is only the instrument in a more profound struggle against opinion, for the misfortune of people comes from opinion' (206). *Thinking*, they argue, must confront chaos (208). Art, science, and philosophy are daughters of chaos, whose other parent may be thought itself (gendering would be difficult, and may be multiple on both sides): 'chaos has three daughters, depending on the place that cuts through it: these are the *Chaoids*—art, science, and philosophy—as forms of thought or creation. We call *Chaoids* the realities produced on the planes that cut through the chaos in different ways' (208).

Neither this vision nor the role of the concept of chaos in it is surprising, given the argument of *What is Philosophy?*. An unexpected and intriguing part of the conclusion, giving it its title, 'From Chaos to the Brain', is a new conception of the *brain*, rather than only the mind, that emerges by an almost sudden shift at this very point. It would not be possible to discuss this extraordinary conception, or its possible future, except by stressing its ultimate grounding in the idea of chaos and of thought, and now the brain, in confrontation with it. This grounding certainly poses new questions about the relationships between mind and the brain, or even mind and matter.

Would our conceptions of each, and of the role of chaos in each, mirror each other? Or would they be subject to different conceptions, thus complicating our picture of either, or their relationships? Would different conceptions of chaos, such as the incomprehensible, chance and disorder, or the virtual, combine differently in our theories? The different ways of confronting chaos by art, science, and philosophy already pose questions concerning their concomitant relations to thought and chaos, even if one leaves the question of the brain aside. Is it the same thought, even if chaos is the same, or are more complex heterogeneity and interactions found already at this level? Perhaps these questions still belong to our thinking as thinking through art, science, and philosophy, and will not be asked by the thought of the future, the thought of the 'people to come'. This type of questioning is where Deleuze and Guattari end their book. But they also look beyond these questions to interferences between the planes of art, science, and philosophy, interference of their wave fields, which takes us to the deeper recesses of thought. As they write:

> The three planes, along with their elements, are irreducible: *plane of immanence of philosophy, plane of composition of art, plane of reference or coordination of science; form of concept, force of sensation, function of knowledge; concepts and conceptual personae; sensations and aesthetic figures, figures and partial observers.* Analogous problems are posed on each plane; in what sense and how is the plane in each case, one or multiple — what unity, what multiplicity? But what to us seem more important now [in approaching the brain] are the problems of interference between the planes that join up in the brain. (216)

The presence of these interferences is essential for our understanding of thought and its confrontation with chaos. In question, however, are not only interferences of art in philosophy, or science in art, or philosophy in science, and so forth. Such interferences are significant, but most crucial are those that, wherever one finds them, are, ultimately, not localizable in any of these three denominations, so defining for our thought and culture now. They are, accordingly, manifestations of that which is still thought, and as such still confronts chaos, but is no longer containable by these denominations. Thus, these interferences manifest a possibility of a different future, perhaps no longer defined by art, science, and philosophy, or their relationships. The future is, as ever, the primary concern of Deleuze's or

Deleuze-Guattari's philosophy, always a philosophy of the future. They write:

> *Philosophy needs a nonphilosophy that comprehends it; it needs a nonphilosophical comprehension just as art needs nonart and science needs nonscience.* They do not need the No as beginning, or as the end in which they would be called upon to disappear by being realized, but at every moment of their becoming or their development. Now, if the three Nos are still distinct in relation to the cerebral plane, they are no longer distinct in relation to the chaos into which the brain plunges. In this submersion it seems that there is extracted from chaos the shadow of the 'people to come' in the form that art, but also philosophy and science, summon forth: mass-people, worlds-people, brain-people, chaos-people — nothinking thought that lodges in the three, like Klee's nonconceptual concept or Kandinsky's internal silence. It is here that concepts, sensations, and functions become indiscernible, as if they shared the same shadow that extends itself across their different nature and constantly accompanies them. (217–18)

These are, then, nonlocalizable interferences that are most crucial for thought, although the localizable ones remain significant, in part as harbingers of nonlocalizable ones. A given work in each domain can manifest such interferences, and for now philosophy, art, and science are the only ways to sense thought that will be defined by and define 'people to come', apart from philosophy, art, or science, but still as a confrontation with chaos. Deleuze and Guattari's concept of the brain points towards this future and the 'people to come' in — or to create — the world without art, science, and philosophy, which, as confrontations with chaos, summon these people forth.

But what kind of thought would it be? What kind of thought could it be? Would 'people to come' ask these questions? Will they ask questions, to begin with? Does thought require questions, as it seems it does for us, or it can struggle with and relate to chaos otherwise, even against opinion, which indeed does not like questions? But will opinion govern the life of the people to come, as it governs ours? It may not be a question of thought, which will struggle with chaos as it has ever done, although for now 'What is thought?' remains a question which, as a confrontation with chaos, is re-posed, along with 'What is the Brain?', by Deleuze and Guattari. It may be a question of what kind of people the 'people to come' will prove to be. A political question, at least for now? But then, such a question cannot only be political, even now.

NOTES

1 Gilles Deleuze and Félix Guattari, *What Is Philosophy?*, translated by Hugh Tomlinson and Graham Burchell (New York, Columbia University Press, 1994), 118. Henceforth *WP*.

2 Throughout this essay 'science' includes mathematics, which is in accord with Deleuze and Guattari's use of the term.

3 Ilya Prigogine and Isabelle Stengers, *Entre le temps et l'éternité* (Paris, Fayard, 1988).

4 Gilles Deleuze and Felix Guattari, *A Thousand Plateaus*, translated by Brian Massumi (Minneapolis, University of Minnesota Press, 1990), 311–15. Henceforth *ATP*.

5 In quantum field theory the effects of Einstein's relativity theory are taken into account, which make the theory relativistic, in contrast to quantum mechanics, where such effects can be neglected because the speed of the objects considered is slow vis-à-vis the speed of light.

6 Frank Wilczek, 'In Search of Symmetry Lost,' *Nature* 433 (20 January 2005), 239.

7 The book was published in English as *Fashionable Nonsense: Postmodern Intellectuals' Abuse of Science* (New York, Picador, 1998). Henceforth *FN*.

8 See Arkady Plotnitsky, *The Knowable and in the Unknowable: Modern Science, Nonclassical Thought, and the 'Two Cultures'* (Ann Arbor, University of Michigan Press, 2002). Specifically on this passage, see 277–8, note 12. The book also offers an extensive discussion of quantum mechanics along the lines of the present article and further references (29–108).

9 See *The Knowable and in the Unknowable* (particularly 204–6 on quantum mechanics and 145–7 on imaginary numbers).

10 See Richard Feynman *QED: The Strange Theory of Light and Matter* (Princeton, Princeton University Press, 1988). This is arguably the best non-technical book on the subject.

11 Gilles Deleuze, *The Logic of Sense*, translated by Mark Lester with Charles Stivale, edited by Constantin V. Boundas (London, Athlone, 1990).

12 See Manuel DeLanda, *Intensive Science and Virtual Philosophy* (London, Continuum, 2002).

13 This concept of chaos as the virtual also suggests a more dynamic, more temporal conceptuality, which may be contrasted with a more spatial view of Deleuze's thought, sometimes argued for by commentators.

14 Deleuze and Guattari also discuss mathematics in terms of 'prospects' (*WP*, 135–62). Prospects, however, are a particular (logical) type of functions, and mathematics is full of other functions, invoked throughout the book.

Chaos and Control: Nanotechnology and the Politics of Emergence

MATTHEW KEARNES

Abstract
This article looks at the strong links between Deleuze's molecular ontology and the fields of complexity and emergence, and argues that Deleuze's work implies a 'philosophy of technology' that is both open and dynamic. Following Simondon and von Uexküll, Deleuze suggests that technical objects are ontologically unstable, and are produced by processes of individuation and self-organization in complex relations with their environment. For Deleuze design is not imposed from without, but emerges from within matter. The fundamental departure for Deleuze, on the basis of such an ontology, is to conceive of *modes of relating to* the evolution of technology. In this way Deleuze, along with Guattari, provides the basis for an ethics and a politics of *becoming* and *emergent control* that constitutes an alternative to the hubris of contemporary reductionist accounts of new areas such as nanotechnology.

Keywords: nanotechnology, emergence, reductionism, evolution, Deleuze

Entia non sunt multiplicanda praeter necessitatem (Ockham's Razor)

Control

Science is part of the cosmos it creates. Contra Descartes, science is not a method or capacity through which we are able to achieve existential externality in order to discover eternal truths. For Bergson the game that science plays in the cosmos is all about time — or in his own terms 'duration'. He states: 'The universe *endures* (. . .) The systems marked off by science *endure* only because they are bound up inseparably with the rest of the universe'.[1] That is to say that the endurance of science is provisional. Its endurance is marked by the universe it describes. This temporal dimension also introduces the possibility for variation and evolution.

Perhaps this is a statement about *S*cience, rather than about *s*cience — a statement that is too all encompassing. We will leave such thoughts for somebody else, for here we are interested in science. We are interested in the specific interventions that particular scientists make, and are making, into the material world within the broadly defined field of nanotechnology. Indeed, it is at this level of specificity that the notion of endurance comes into its own. For nanotechnology, which is both scientific and technical (if we must bring up that old distinction), is concerned with making things. That is, nanotechnology is defined by the construction, generation and growth of objects, devices and architecture — all of which have a certain (provisional) endurance. In working at the nanoscale (10^{-9}m), in the world of Brownian motion and atomic uncertainty, this kind of endurance is produced by certain forms of control — specifically the control of sub-molecular particles, of biological systems, chemical syntheses, reactions and crystal growth. If it is possible to construct a nanostructure from a few atoms or molecules, or to grow one using a protein or some process of crystallization, the endurance of this structure is dependent on being able to control atomic-level forces that would tear it apart. Such endurance is premised on perpetual control.

Of course Deleuze was also famously interested in control, particularly in his 'Postscript for Control Societies'. For Deleuze, 'control' defines the political constitution of the contemporary moment. Critical of Foucault's analysis of modern discipline, he suggests that institutions and technologies of incarceration and discipline are being replaced by the mechanisms of control. He states: 'We're in the midst of a general breakdown of all sites of confinement — prisons, hospitals, factories, schools, the family (. . .) *Control societies* are taking over from disciplinary societies'. For Deleuze, this regime of control is not defined by any specific mechanism, technology or institution. Indeed, he states:

> It's not a question of amazing pharmaceutical products, nuclear technology, and genetic engineering, even though these will play their part in the process. It's not a question of asking whether the old or new system is harsher or more bearable, because there's a conflict in each between the ways they free and enslave us. With the breakdown of the hospital as a site of confinement, for instance, community psychiatry, day hospitals, and home care initially presented new freedoms, while at the same time contributing to mechanisms of control as rigorous as the harshest confinement.[2]

This then is one side of control: the side of power and determinism. This is the power of *total* control and it is the dream of many nanotechnologists. This is the kind of control through which some suggest that it will be possible to 'build anything we want' simply by arranging atoms the way we would like them.[3] However, for Deleuze, control is never absolute in this sense. Control is a product of a repetition of force. Therefore in the application of force and control we also see the radical possibility for creativity and escape. As such Deleuze is more interested in modes of control and modes of perpetuating coherence than in absolute control per se.

Deleuze's 'philosophy of technology' is both open and dynamic. He adopts an open *stance* in relation to science and technology. For Deleuze, the link between 'basic' scientific knowledge and technical systems — which is mediated through the disciplines of engineering, design, predictability and control — is neither simple nor necessary. For example, Deleuze's critique of the hylomorphic schema suggests instead that control and predictability emerge almost spontaneously. As such, and following Simondon and von Uexküll, technical objects for Deleuze are ontologically unstable, produced through processes of individuation and self-organization in complex relations with particular milieus.[4]

What is significant here is not the scientism of Deleuze's own thought, but rather the stance that Deleuzian thought enables one to take in relation to the emergence of technical objects and systems. Recent scholarship on the use of scientific concepts in Deleuzian thought has tended to be polarized between realist and metaphorical interpretations: approaches that defend his use of mathematics and physics as either scientifically valid or allegorically salient.[5] Both of these positions interpret the use of science by Deleuze in solely conceptual terms — as if what is at stake is *either* the metaphysical rigour of his use of science or its metaphorical resonance. However, this choice between realism and idealism is a false one. Both (mis)understand science itself in *wholly* conceptual terms without any sense of the interconnected material practices and interventions fundamentally intertwined with the emergence of technical objects and systems.[6] Deleuze's use of science is much more political than philosophical, and much more attuned to the mechanisms of invention and creation to be simply cast as conceptual intellectual folly.

What is at stake in Deleuze is not Science, or even Philosophy. This is particularly the case in Deleuze and Guattari's almost 'disrespectful' treatment of the pillars of Science, Art and Philosophy in *What is Philosophy?*, where they state:

> The three disciplines (art, science and philosophy) advance by crises or shocks in different ways and in each case it is their succession that makes it possible to speak of 'progress'. It is as if the *struggle against chaos* does not take place without an affinity with the enemy, because another struggle develops and takes on more importance — the struggle *against opinion*, which claims to protect us from chaos itself.[7]

Deleuze and Guattari's interpretation of science is fundamentally materialist. The 'affinity with the enemy' is the same relation that the artisan makes with the wood, or the blacksmith makes with the metal outlined in *A Thousand Plateaus*.[8] In this sense, Deleuze and Guattari refuse the Cartesian definition of Science as simply conceptual invention. Rather, for Deleuze conceptual science is one expression of this material 'affinity with the enemy', in the same lineage as alchemy, woodwork and blacksmithery. To this end, in what follows I will outline Deleuze's stance in relation to invention, creation and technology. I will suggest that, in place of the reductionism of 'total' control, Deleuze's ethic aims to vitalize technology. Deleuze aims to open up and potentialize science and technology to the internal evolution of matter 'all the way down'.

Nanotechnologies

The prefix *nano*, from the Greek *nanos* meaning small, gives an immediate indication of the kinds of interventions in the material word envisioned by this word — small ones. A nanometre being 10^{-9}m, nanotechnology is technology on the atomic and submolecular scale. Nanotechnology encompasses work of advanced nanoscale science, particularly the increased understandings of atomic-scale interactions and the capacity to visualize (or more correctly to characterize) and control materials at sub-micron levels using the *scanning tunnelling microscope*. However, as suggested by the suffix — *technology* — nanotechnology is also a term that designates new forms of practice at nano-metre scale.

The canonical story of the origin of nanotechnology is familiar and oft told. The Nobel Prize winning physicist Richard Feynman's now famous lecture, 'There is Plenty of Room at the Bottom' is commonly regarded as the first public musings by a scientist about the possibilities of technology on the nano-scale.[9] His notion that there is no physical barrier to the extreme miniaturization of technology operates as a central motivating discourse around which nanotechnology operates. Indeed in subsequent controversies around what counts as 'fact' and 'fancy' proponents have often claimed that they are simply expressing the implications of Feynman's original vision. One example is Eric Drexler's now (in)famous nanotechnology manifesto, *Engines of Creation: The Coming Age of Nanotechnology*, in which he outlines a Feynmanian notion of the sheer physical possibility of molecular nanotechnology as an alternative to modern 'bulk technology':

> Coal and diamonds, sand and computer chips, cancer and healthy tissue: throughout history, variations in the arrangement of atoms have distinguished the cheap from the cherished, the diseased from the healthy. Arranged one way, atoms make up soil, air, and water; arranged another, they make up ripe strawberries. Arranged one way, they make up homes and fresh air; arranged another, they make up ash and smoke.[10]

For Drexler, nanotechnology represents a mode through which life may be understood *physically*. Drexler's manifesto mirrors Feynman's vision: that there is no physical barrier or practical reason why current technology — medicine, information technology and engineering — cannot operate with precision at the nano-scale.

Nanotechnology is fundamentally technological. Its main objectives are forms of technique, process, and precision. Drexler's manifesto, which self-consciously mirrors science fiction in developing possible future nanotechnology scenarios, does not break new ground in terms of scientific knowledge regarding the atom. Rather it is a preparatory exploration, outlining the possible implications of the convergence of abstract atomic knowledge with increased technical ability at these scales.

Nanotechnology and 'reductionist returns'

The reductionism implicit in contemporary genetic technologies[11] is both extended and intensified in Drexler's account of nanotechnology.[12] For Drexler all things, both organic and inorganic, are simply a collection of atoms and molecules. In this sense Drexler's determinism is fundamentally physicalist. Whereas for genetic determinism it is assumed that life forms are determined by the *process* of heredity, for Drexler life itself is determined simply by its physical constitution. In this sense, life is absolutely divisible, and therefore manipulable. The only difference between material objects is the alternative arrangements that atoms take. Indeed, the radical possibility that Drexler presents is that, when control over the structure of matter is achieved, it will be possible to make 'almost anything' from the 'bottom-up'.[13]

The root of this physicalist determinism is Schrödinger's essay *What is Life? The Physical Aspect of the Living Cell.* In this essay Schrödinger outlines a physicalist understanding of life and matter upon which the dreams of an unlimited material abundance, produced by nanotechnology, are based. Indeed he initiates a 'materialist turn' in biology by suggesting that biological processes may be explained physically. Schrödinger asks: 'How can events in space and time which take place within the spatial boundary of a living organism be accounted for by physics and chemistry?' To which he gave the following answer: 'The most essential part of a living cell — the chromosome fibre — may suitably be called *an aperiodic crystal*.'[14] Schrödinger signals the fundamental idea that living cells are physical — that they are composed of physical elements, atoms and aperiodic crystals — and as such can be conceptualized in physics and chemistry as well as biology. The notion, articulated by proponents of nanotechnology, that such technologies may be able to cure diseases, remedy pollution and create clean energy, relies on this most basic premise of the atomic physicality of 'life'. Therefore Schrödinger's physicalism is paradoxical in that it also heralds a reductionist notion of the absolute divisibility of all life. In this sense the physical itself becomes nothing less than an instrumental concern in the technologization of life and nature.[15]

There is a rich double meaning to Schrödinger's essay *What is Life?* His notion of matter and of life is irreducibly vitalistic. He is interested in *life* — in the physics of life. In one sense this life is about the physicality of life, its material and atomic bases, yet in another it is about a certain physical life or liveliness of matter. Schrödinger introduces a notion of the vital life of the atomic. He captures the movement, variation and digression of the material at atomic and molecular scales. Crandall embraces Schrödinger's move in his vision of nanotechnology: 'This bumbling, stumbling dance allows molecules all possible "mating configurations" with the other molecules in their local environment. By variously constraining and controlling the chaos of such wild interactions, biological systems generate the event we call life.'[16] For Crandall the *promise* of nanotechnology is of a technology of 'molecular construction', the miniaturization of macro-scale manufacturing techniques enabling the precise placement construction of objects atom by atom. Similarly, Drexler's vision is of nano-scale replicators — autonomous, self-replicating machines will enable such techniques of atomic precision to be infinitely multiplied. Building upon enhancements in the characterization of atomic structures and the precision with which these structures may be controlled, Crandall positions nanotechnology as the ability to create nano-scale *machines* and to construct material objects from the bottom up, through the precise alignment of sub-molecular materials.

For both Crandall and Drexler the simple fact that biological life is accomplished through the selective control over the movement of atoms and molecules demonstrates the possibility of similar human-designed processes. Their ontology of the atomic, though emphasizing the internal movement, variation and flux, operates only as the technical limit upon what is possible. For Drexler and Crandall, the technical limits of nanotechnology are 'set' by the nature and being of atomic scale matter (its ontological status). At the core of the Drexlerian vision, as expressed by Crandall, is a notion of control. If the 'event we call life' is generated by the 'wild interactions' of particles at the nanoscale, the broad goal of fundamental nanoscale research must be toward achieving control over these interactions so that they may be directed in desired ways. As such, the Drexlerian vision of the *possibilities* for self-replicating nano-scale robots has become a programmatic definition of nanotechnology as the 'total (or near total) control over the structure of matter'.[17]

The mechanistic reductionism of nanotechnology also has its roots in a logic of what might be properly termed a 'biological turn' in theoretical physics and mathematics. This 'biological turn' may be identified in the move toward the mathematical modelling of complex and naturally occurring phenomena — particularly in swarm and game theory. Of particular significance in the development of nanoscience and nanotechnology is von Neumann's mathematical modelling of self-reproducing systems. Von Neumann's theory of automata — modelled on the functionality of the neuron — is an expression of the sheer algorithmic possibility of computationally recreating naturally occurring self-reproducing systems. His is a vision of what is 'logically possible': 'A new, essentially logical, theory is called for in order to understand high-complication automata and in particular the central nervous system.'[18] As such, von Neumann's thesis suggests that naturally occurring self-replicating systems may be modelled mathematically. For von Neumann, the key to the mathematical recreation of cellular automata is the inherent logic embedded within the very complexity of such systems — what he terms 'complication'. At a high degree of complication — or complexity — systems are able to self-organize themselves. The key therefore is to discover the mathematical logic and laws through which systems of such high complexity operate:

> All these are very crude steps in the direction of a systematic theory of automata. They represent, in addition, only one particular direction. This is, as I indicated before, the direction towards forming a rigorous concept of what constitutes 'complication' (or complexity). They illustrate that 'complication' on its lower levels is probably degenerative, that is, that every automaton that can produce other automata will only be able to produce less complicated ones. There is however a certain minimum level where this degenerative characteristic ceases to be universal. At this point automata which can reproduce themselves, or even construct higher entities, become possible.[19]

For von Neumann, 'self-reproduction, evolution — *life* in brief — can be achieved within a cellular automaton — a toy world governed by simple discrete rules not unlike those of a solitaire game'.[20] Von Neumann's search was for the laws that governed the formation and functioning of cellular automata. His logical route is from the big to the small, from the complexity of self-organization to simple, discrete laws. Von Neumann offers a bio-mimetic logic where existing biological systems may be modelled through such laws and the precise control of the parameters of such systems. As such, von Neumann's notion of the algorithmic recreation of complex systems masks an extreme reductionism in which simple laws control complex systems.[21]

The importance of Von Neumann's work in nanoscience and nanotechnology is two-fold. Firstly, his basic thesis that self-replicating automata are *logically* possible casts matter and the material as simple instrumental concerns. For von Neumann — and also for Drexler — if self-replicating automata are logically possible they must also, by necessity, be physically possible. Secondly, the combined effect of Schrödinger's physical understanding of life and von Neumann's logic of the self-replicating automata is to suggest that chemical and biological processes may be understood functionally as code or information. Von Neumann suggests that naturally occurring automata conform to algorithmic logic and that they may be technically recreated by manipulating the parameters of algorithms. This reductionism offers the possibility of total control of such biological and chemical systems. In this way biological systems become a set of mathematical and computational instructions that may be technologically re-ordered. Lehn echoes this understanding of biological and chemical processes by suggesting that forms of 'living' or 'complex' matter may be created by controlling the informational exchange at the heart of chemistry and biology. In what he terms 'supramolecular' chemistry, he suggests:

> Supramolecular chemistry has paved the way toward apprehending chemistry as an information science through the implementation of the concept of molecular information with the aim of gaining progressive control over the spatial (structural) and temporal (dynamic) features of matter and over its complexification through self-organisation, the drive to life.[22]

Both the 'materialist turn' in biology and the 'biological turn' in mathematics and physics are concerned with a set of logical possibilities. Schrödinger's conception of the physics of life and von Neumann's mathematical theory of automata have the effect of converting life itself into discrete physical entities which operate as a form of information or code. This double move has the paradoxical effect of rendering the physicality of biological and chemical systems as merely instrumental concerns in the hylomorphic application of computational models onto material substrates. Combined with Feynman's vision of atomic-scale machinery, nanotechnology operates as a set of theoretical promises and possibilities for gaining 'progressive control' over the structure of matter in the design and manufacture of nanotechnologies. Life itself is cast as absolutely divisible. The mechanisms of reproduction and self-organization are themselves recreatable, given the precise control over the parameters of chemical synthesis of biological systems.

This rhetorical move from the big to the small, from the complex to the simple, and from the chaotic to the organized, parallels the overall imagination of nanotechnology as the ability to precisely control the ultimate building blocks of life. This logic is also comparable to the reductionism at the heart of some versions of complexity theory. Broadly speaking, whereas chaos theory works in the reverse direction — small events producing large results — complexity theory suggests that simple structures emerge and self-organize in the context of complex and dynamic systems.[23] Although the spontaneous emergence of organized structures is inherent to theoretical accounts of complexity theory, its technological operationalization often reveals a reductionist drive toward simplification and predictive control.[24] It is in the construction of complexity theory as a unifying project, through which total systems understanding, simplification and predictive control may be achieved, that complexity theory is at its most reductionist. For example, Capra defines complexity theory as: 'A new mathematical language and a new set of concepts for describing and *modelling* complex nonlinear systems. Complexity theory now offers the exciting possibility of developing a *unified* view of life by integrating life's biological, cognitive and social dimensions' (emphasis added).[25] By unifying biological, chemical and physical knowledge, complexity theory is thought to enable an enhanced capacity to model non-linear systems. By extension, complexity theory is seen to enable the precise control and recreation of such systems.[26] The rhetorical move from the complex, the large, and the extensive to the simple, the small and the intensive is ambiguously reductionist. Given this ambiguity, complexity theory masks a 'reductionist return'[27] in contemporary *techno*science that is revealed in the currency of notions such as predictive control, modelling, law and total systems knowledge.

This reductionism mirrors the rhetorical efforts of miniaturization and simplification made by Feynman, Schrödinger and von Neumann. The combined effect of Feynman, Schrödinger and von Neumann is to cast biological, chemical and material 'life' as absolutely physically divisible and created through mechanisms that are ultimately controllable. Indeed, nanotechnology operates as a 'unifying' project similar to that of complexity theory — combining the traditional scientific disciplines of physics, chemistry, mathematics and biology with the technically-oriented disciplines of engineering and computing. Rhetorically, nanotechnology also relies upon a similar rhetorical move from the big to the small as the ultimate technical expression of the miniaturization imperative. Thus for Drexler, following Feynman, Schrödinger and von Neumann, the sheer logical possibility of nanoscale engineering and manufacture establishes the absolute reducibility of all forms of life and materiality to the atom and the technical possibilities for building things 'atom by atom'.

Unity versus singularity

The reductionism of advanced nanotechnology is also deeply political. The vision of nanotechnology as heralding the ability to remake the world 'atom by atom', and as leading to the 'next industrial revolution', is also a State-sanctioned vision of the power of science to revolutionize material practice.[28] The reductionism of nanotechnology, which demands total control of the atomic scale, is deeply entwined with this politics. This is what Deleuze calls the politics of the State, or the 'apparatus of capture', in which the unifying project of reductionist science works toward the total control demanded by the State.

Deleuze's ontology is of an entirely different order. Deleuze neither moves from the complex to the simple, nor from the simple to the chaotic. Rather he starts with the singular or — more properly — the singularity. Whilst in nanotechnology the unity represents the absolute divisibility of life, Deleuze starts with the notion of the singularity as the basis for molecular variation and flux. His 'philosophy of difference' is concerned with revalorizing the singular, over and above the particular. Consequently, he deploys an explicitly monistic ontology — a material pantheism whereby the singularity of matter is alive with the creative potential of endless evolutions and innovations. Deleuze states:

> There has only ever been one ontological proposition: Being is univocal. (. . .) There are not two 'paths', as Parmenides' poem suggests, but a single 'voice' of Being which includes all its modes, including the most diverse, the most varied, the most differenciated. Being is said in a single and same sense of everything of which it is said, but that of which it is said differs: it is said of difference itself.[29]

What Deleuze does here is to free 'the singular' from 'the particular', giving it an individuating capacity. His notion of singularity is at once an absolute rebuttal of both the Platonic and Aristotelian metaphysics of matter and a valorization of the creative vitalism of the material. He refuses the categorical difference, established by the metaphysicians, between matter and form or between the subject and the object. Rather, all things are *formed* through repetitious individuation of the same substance — the monadic singularity — intensities, riffs, sublimations in a singular key. Rethinking monadology in explicitly materialistic terms enables Deleuze to insist upon a materialism that is 'roughly equivalent to an ongoing Big Bang, permanent Creation',[30] precisely because whilst this evolution is both permanent and multiple, the substance upon which these operations are performed is singular. Thus it is not simply that 'matter is singular' as a universal substance. Rather, matters are singularities, momentary agglomerations in the creative evolution of the singular, monistic substance.

As a consequence of his emphasis on a monistic creative evolution, Deleuze's materialism is inherently spontaneous. For Deleuze, monism is not a form of reductionism to the atomic, the whole, the plane, or the easy to handle. Rather, in Deleuze's hands monism results in the elevation of the singular, the singular that is difference itself. Deleuze's philosophy of difference is singular precisely because it positions difference as internal to the object, rather than between (categorically different) things. In this way, he follows Simondon's drive to free *individuation* from any organizing principle of the *individual*.[31] Deleuze's aim is to compose a philosophy of difference — rather than of diversity — in the singular, where what differs is not one thing from another, but the thing from itself. This repetition of the singular, what Deleuze calls the production of singularities, imbues objects, things, substances and bodies with a dynamic sense of action. For Deleuze this spontaneity goes 'all the way down' and is not computable by law or assumptions of predictability. It is by following Simondon that Deleuze enables a dynamic theory of technology that resonates with the technical possibilities of nanoscience, in which the singular, the atomic and the molecular are energized in the creative production of difference. Repetition, for Deleuze, is the essence of creativity or — in Bergson's terminology — creative evolution. It is this monistic ontology of singularity (rather than unity) that imbues matter with a sense of unstable movement because the *repetition* of this monistic substance is not simply a matter of the production of equivalences. Rather, repetition is the production of difference, the movement and creative evolution of the thing: 'Repetition is a condition of action before it is a concept of reflection. We produce something new only on condition that we repeat — once in the mode which constitutes the past, and once more in the present of metamorphosis.'[32]

Though Deleuze does not have what might be termed a 'philosophy of technology' — in the manner of Heidegger or even Derrida — one would imagine a Deleuzian stance or ethic *toward* technology that, following Simondon and von Uexküll, allows for the creative individuation of technology. Deleuze's stance toward technology is fundamentally a political engagement with powers of invention and creation. His basic concern is to free individuation, and the singularity, from the unifying project of 'total control'. Whereas in nanotechnology — as in other contemporary technologies — it is imagined that internal deviance, evolution and spontaneous self-organization may either be mastered or technically harnessed, Deleuze imagines a technology that is radically open to evolution 'all the way down'. Indeed, the problem for Deleuze with philosophies of technology (either the sceptical, Heideggerian versions of technology as threat,[33] or the more positive endorsements of the social-evolutionists[34]) is that they treat technology as a distinct ontological category. Deleuze's philosophy is more an attitude or stance toward technologies that is open to the internal flux of technology, down to the molecular level.

This stance toward technology imbues Deleuze's attitude toward the 'total control' imagined in control societies. For Deleuze, control is not symptomatic, but rather emblematic. It does not emanate from a particular technology, or the 'technologization' of all forms of life itself, but is rather a kind of contemporary epistemic moment. The kind of control that Deleuze speaks of is the perpetual control of noise, variation and flux — the same modes of control over the internal differentiation of matter imagined in nanotechnology. However, this control is expressed in specific modes of control. Because, for Deleuze, control necessitates the minute control of flux and variation, it is itself perpetually changing and never complete. For example, he states: 'Control is short-term and rapidly shifting, but at the same time continuous and unbounded.'[35]

Nanotechnology and the mastery of evolution

Flux and variation are also of concern to nanotechnologists. Much nanotechnology is defined by the attempt to limit and control the evolution of biological and chemical systems. For example, Drexler's vision of nanotechnology represents an inherently mechanical mode of mastering atomic-scale flux. He seeks to both control and harness the movement, variation and digression of matter at atomic and molecular scales. Drexler's vision is largely mechanistic, suggesting that nanotechnologies can be produced through the hylomorphic imposition of an external design on a material substrate, through the precise manipulation of matter 'atom-by-atom'. His basic premise, following Feynman, is that there is no physical impediment to conceiving manufacturing technologies on the atomic scale. He represents this vision as a miniaturization of existing mechanical techniques to the atomic scale. Indeed, this mechanical bias of his thesis has been the source of a number of significant criticisms, primarily that the precise mechanical precision required is simply impossible.[36] Despite the extensive criticism of Drexler, both his goal of a 'bottom-up molecular project' and the broad project of gaining 'control over the structure of matter' maintain an important rhetoric in the field. In the extensive criticism of Drexler by Richard Smalley,[37] and latterly Richard Jones,[38] this notion of control is actually intensified. What is at issue for Smalley is whether the precise control of atomic structures, necessitated by Drexler's vision of nano-scale manufacture, is technically possible.

Due in part to the repudiation of the vision of 'molecular manufacture', Drexler's radically mechanistic version of nanotechnology has been substituted by a more conservative and pragmatic set of nanoscale possibilities.[39] At issue is the sheer physical possibility of autonomous nanoscale machines and the precision necessary to 'create' them. Recent scholarship has, however, revived the radical possibility of nanoscale machinic autonomy by rethinking the very way in which such machines might be created. In this biomimetic model, such machines are grown using existing biological systems as working examples. In the words of Bernadette Bensaude-Vincent, there are 'two cultures of nanotechnology'.[40] Both present different understandings of the relationship between *design* and matter. One version of nanotechnology is implicitly mechanical. The material world is completely atomized as simply an aggregate of *particles*. It is in this sense that it is suggested that it will be possible to create anything 'from the bottom-up', simply by arranging atoms and molecules in certain (desired) ways. Alternatively, Richard Jones imagines a different form of nano-scale control in which technical interventions *harness* rather than *master* the chaotic interactions at the nanoscale. He re-presents these same goals, but changes the route through which they will be achieved. He suggests that nano-scale machines will be achieved through a more bio-mimetic, or emulatory, nanotechnology which both takes its inspiration from, and actively utilizes, existing biological systems.[41] Rather than imagine nanotechnology as a set of mechanical interventions at the nanoscale, Jones presents a vision of nanotechnology that is modelled on naturally occurring biological systems. For Jones, naturally occurring 'molecular assemblages' such as protein and DNA represent a functional equivalent of self-assembling and self-organizing molecular machines. Jones' thesis is that biological systems present functioning models through which more purposeful nanotechnological interventions may be made. For Jones, bionanotechnology is a route through which applications at the nanoscale may be achieved. He states his basic vision as follows:

> My own view is that radical nanotechnology will be developed, but not necessarily along the path proposed by Drexler. I accept the force of the argument that biology gives us proof in principle that a radical nanotechnology, in which machines of molecular scale manipulate matter and energy with great precision, can exist. But this argument also shows that there may be more than one way of reaching the goal of radical nanotechnology, and that the path proposed by Drexler may not be the best one to follow.[42]

Significantly for Jones, this necessitates a design process that is both open to, and able to harness, molecular evolution: 'Evolution needs some kind of selection pressure — some kind of way of deciding which of the many random changes in the molecular sequence should survive and prosper.'[43] Jones maintains that nanoscale machines are both technically possible and a desirable route through which wide technical advances may be made, but suggests that the current 'engineering approach' may well prove unrealistic. Instead he adopts a bio-mimetic approach, suggesting that the creation of molecular machines, necessary to fulfil the radical vision of 'bottom up' manufacture, may be created by copying nature — that is emulating existing self-replicating systems such as protein and DNA. Similarly, Seeman and Belcher outline this aim of 'emulating biology' by re-creating self-assembling systems:

> A key property of biological nanostructures is molecular recognition, leading to self-assembly and the templating of atomic and molecular structures. For example, it is well known that two complementary strands of DNA will pair to form a double helix. DNA illustrates two features of self-assembly. The molecules have a strong affinity for each other and they form a predictable structure when they associate. Those who wish to create defined nanostructures would like to develop systems that emulate this behaviour. Thus, rather than milling down from the macroscopic level, using tools of greater and greater precision (and probably cost), they would like to build nanoconstructs from the bottom up, starting with chemical systems.[44]

Although there is a strong deterministic logic central to the representation of nanotechnology,[45] this sense of momentary control describes the practice of actually building nanostructures. For example, Karen J. Elder discusses polymerization as a way in which nano-scale structures may be constructed.[46] This is a curious model of a kind of construction where a nano-scale architecture is *grown* rather than accomplished. The resulting form, though desired, is perhaps temporary. Indeed, the object of such research is to control the growth in such a manner that it evolves in desired ways. Similarly, take Seeman and Belcher's description of the use of biological systems in building complex and functional systems:

> In natural systems, macromolecules exert exceptional control over inorganic nucleation, phase stabilisation assembly, and pattern formation. Biological systems assemble nanoscale building blocks into complex and functionally sophisticated structures with high perfection, controlled size, and compositional uniformity. (. . .) The exquisite selectivity of complementary biological molecules offers a

possible avenue to control the formation of complex structures based on inorganic building blocks such as metal or semiconductor nanoparticles.[47]

This notion of construction entails capitalizing on the interactions of biological systems — in this case rDNA, DNA and protein — in order to create desired patterns and objects. The object here is also control, but a control that is both provisional and active. Both of these design paradigms, though designed to achieve control, are in fact more like specific modes of control. In each, desired constructions — shapes, geometries and architectures — are the product of specific and perpetual control. Indeed, what is entailed is the control of interactions and processes in which we always sense, as Deleuze suggests, the possibility of escape.

Both Drexler's mechanistic vision of nanotechnology and the more biologically nuanced vision of the bio-mimeticists create different relations with evolution. Both are ways of 'relating to' the flux caused by the internal evolution and variation of objects at the molecular scale and technical design paradigms in this context. Both cultures are ways of harnessing and (more significantly) limiting the creative force of evolution. Instead of imagining the possibility of simply arranging atoms mechanically, biomimetic nanotechnology aims to capitalize on the functionality of existing biological systems, particularly protein and DNA.

However, the distinction between these two versions of nanotechnology is never strictly defined as, in a sense, both are concerned with designing nano-scale mechanisms and controlling their *operation*. For mechanistic nanotechnology it is necessary to control the 'wild' Brownian interactions of atoms and molecules in order to create stable and functional objects. Similarly, bio-mimetic nanotechnology requires the precise control of biological systems in order to achieve desired and designed outcomes. Despite the hubris of complete design in nanotechnology, these kinds of nano-scale interventions have produced only temporary states of order.[48] Whereas in mechanistic versions of nanotechnology it is imagined that the complex relations of atomic and molecular particles — the internal movement of matter that is the equivalent of Deleuze's machinic phylum — will be *overcome* by the disciplines of design and engineering, in biomimetic nanotechnology it is suggested that technical objects will be created by utilizing this very complexity. In particular it is suggested that nanotechnologies will be created by piggybacking on existing self-replicating systems such as protein, DNA or rDNA.

At issue then is not simply a Deleuzian openness, or affirmation, of flux, variation and evolution. Indeed, whereas Deleuze appropriates Simondon's understanding of the spontaneous individuation of technology — producing a dynamic theory of technology[49] — both 'cultures' of nanotechnology seek to limit this anarchic potential for evolution 'all the way down'. Indeed, though biomimetic nanotechnology seeks to utilize existing biological machines, and artificially evolve new forms of such machinery, the real work of such a design paradigm is ensuring the stability of the resulting structures and systems. Take for example, Michael Conrad's model of 'emergent computation through self-assembly'.[50] He presents a framework in which the 'self-organising capacities of biological systems are extended expressions of nonlinearity inherent in the time evolution of the universe'.[51] Biological systems are the emergent, and hence spontaneous, effects of non-linear processes. Conrad suggest that such machines are internally fluctuating, challenging the ways that such systems may be designed or harnessed technically. He states:

> If a collection of components is allowed to self-organise in the first place (. . .) then self-consistency is automatically ensured. In general such self-organised aggregates do not perform functions desired by the investigator observing their formation, unlike the totally macroscopic machines that are pasted together in a planned way by a designer with a definite conception in mind. However, the adaptivity of self-organising systems allows for moulding of the function in a step by step trial and error fashion, just as biological organisations adapt through step by step variation and selection. (. . .) Simple reverse engineering of existing biological organisations cannot work, according to the fluctuation model, since they ignore the hysteretic properties of the vacuum sea.[52]

Conrad's model is of a moulding of the functions of biological systems, by trial and error. Conrad re-imagines the place of design — 'reverse engineering cannot work' — insisting on a level of artisanal intimacy in the creation of human-directed biological machines. The incompleteness of the technical control over the atomic scale necessitates, for Conrad, a more open-ended conception of the technological possibilities at this scale.

Deleuze's evolutionary ethic

Such *inventions* are provisional stabilizations of processes of chemical synthesis and biological interaction. This is inconsistent with the vision of nanotechnology as being able to 'create anything from the bottom up'. Rather, such inventions necessitate a Deleuzian sense of the temporality of control and the individuation of technical systems. It is in this sense that Deleuze's understanding of the dynamism of control finds its material expression. Similarly, Oyama develops a notion of the 'emergence of control':

> Control of development and behaviour may be said to emerge in at least three senses. First, it emerges in interaction, defined by the mutual selectivity of interactants. Second, it emerges through hierarchical levels, in the sense that entities or processes at one level interact to give rise to the entities or processes at the next, while upper-level processes can in turn be reflected in lower-level ones. Third, control emerges through time, sometimes being transferred from one process to another.[53]

The crucial point for Oyama, and for Deleuze, is that control *emerges*. It emerges in interactions with the very processes that are (to be) controlled. This means both that control is partial and that design itself emerges with the object. The word emergence here is crucial. In Seeman and Belcher's constructions of nano-geometries using DNA, we might speak of the emergent self-organization of such structures. The design, control and precision necessary to generate such nanoscale geometries emerges with the structures themselves and the processes through which they are formed.

The importance, for Deleuze, of the emergence of control is that it signals the ontological incompleteness of design and the spontaneous variation of technology. This, for Deleuze, is both ontological and ethical and defines his own particular stance in relation to evolution and variation. Deleuze draws strongly on Bergson's notion of variation in developing this ethic. He finds in Bergson the idea of the constant repetition of the object (its simulacrum) that suggests that the object is opened up, repotentialized, returned as constantly differentiated, and constantly multiple. In Deleuzian terms, repetition introduces a form of differentiation that replaces the term IS in the being of the object (X = X = NOT Y) with AND (... X AND Y AND Z AND...): 'Substitute the AND for IS. A *and* B. (...) The multiple is no longer an adjective which is subordinate to the One which divides or the Being which encompasses it. It is a noun, a multiplicity which constantly inhabits each thing.'[54]

Bergson offers a complex and simultaneous analysis of both the 'thing-in-itself' and the thing as an 'aggregate of images'. Deleuze sees in Bergson a fundamental critique of the notion of categorical difference:

> If philosophy is to have a positive and direct relation with things, it is only to the extent that it claims to grasp the thing in itself in what it is, in its difference from all that it is not, which is to say in its *internal difference*.[55]

Bergson's concept of variance becomes crucial. It is a fundamental critique of the notion of categorical difference. Rather than posit a fundamental or essential difference between objects, Bergson allows difference itself to be something. Objects differ from themselves — internally. This is not to obliterate distinction but to fundamentally reconceive the notion of natural difference. Differentiation acts, according to both Bergson and Deleuze, not between objects, but within objects. For both Bergson and Deleuze matter is anything but a boundary. Rather matter is internally unstable.

This material waywardness is not simply metaphorical, or a kind of characterization of matter. Rather it is a waywardness that is based in Deleuze's molecular ontology, and the internal flux of matter at that scale. It is also inherently personal and ethical, defined by an openness to change and variation, all the way down to the molecular level. Take, for example, Deleuze and Guattari's description 'becoming-dog':

> Do not imitate a dog, but make your organism enter into composition with *something else* in such a way that the particles emitted from the aggregate thus composed will be canine as a function of the relation of movement and rest, or of molecular proximity, into which they enter. Clearly, this something else can be quite varied, and be more or less directly related to the animal in question.[56]

Becoming, in this sense (though often misrepresented) is a fundamentally materialist ethic, open to internal variation. In Deleuze's ontology this 'something else' is always difference itself: the action of variation and flux, in an ethic that enters into a relation with difference itself. It is a kind of atomic flux between the object/subject (of course these terms no longer mean much) and *the other*. It is precisely these same internal variations that are the subject of both design and construction in nanotechnology. Deleuze suggests that, because of such self-differentiating movement and variation, design and construction are more complex than admitted in the hubristic accounts of nanotechnology. For Deleuze design is not imposed from without, but emerges from within matter. The fundamental departure for Deleuze on the basis of such an ontology, is to conceive of *modes of relating to* the evolution of technology. For Deleuze and Guattari this ethic — or even politics — 'is a question of arraying oneself in an open space, of holding space, of maintaining the possibility of springing up at any point: the movement is not from one point to another, but becomes perpetual, without aim or destination, without departure or arrival'.[57]

Such an ethics adopts an 'artisanal' stance similar to that of Conrad's design by trial and error. For Deleuze and Guattari, it is a question of 'arraying oneself in an open space' in the same way that Conrad imagines a trial-and-error design process. Finally then, this Deleuzian stance, or attitude, is open to the paradox of the individuating endurance of both science and technology, and to possibilities for spontaneous eruption in these temporal arrangements.

Acknowledgements

The preparation of this paper was supported in part by an ESRC grant (RES-338-25-0006) and a doctoral fellowship at The University of Newcastle. Thanks to Bulent Diken, Adrian Mackenzie, Bronislaw Szerszynski and Brian Wynne for productive conversations in the preparation of this paper.

NOTES

1 Henri Bergson, *Creative Evolution*, translated by Arthur Mitchell (Dover Publications, New York, 1911 [1988]), 11.
2 Gilles Deleuze, 'Postscript on Control Societies', in Gilles Deleuze, *Negotiations: 1972–1990*, translated by Martin Joughin (New York, Columbia University Press, 1995), 178.
3 See K. Eric Drexler, *Engines of Creation: The Coming Era of Nanotechnology* (Anchor Books, New York, 1986); Neil Gershenfeld, *FAB: The Coming Revolution on Your Desktop — From Personal Computers to Personal Fabrication* (New York, Basic Books, 2005); Ray Kurzweil, *The Singularity is Near: When Humans Transcend Biology* (New York, Viking, 2005).
4 See Adrian Mackenzie, *Transductions: Bodies and Machines at Speed* (London, Continuum, 2002).
5 See DeLanda's strong realist interpretation and defence of Deleuzian appropriation of concepts from mathematics and physics in *Intensive Science and Virtual Philosophy* (London, Continuum, 2002). This position is in broad agreement with that taken by Protevi and by Bonta and Protevi in John Protevi, *Political Physics: Deleuze, Derrida and the Body Politic* (London, Athlone, 2001);

and Mark Bonta and John Protevi, *Deleuze and Geophilosophy: A Guide and Glossary* (Edinburgh, Edinburgh University Press, 2004). Alternatively James Williams interprets Deleuze's use of scientific concepts as purely metaphorical, in James Williams, *Gilles Deleuze's Difference and Repetition: A Critical Introduction and Guide* (Edinburgh, Edinburgh University Press, 2003).

6 See for example, Isabelle Stengers, *Power and Invention: Situating Science* (Minneapolis, University of Minnesota Press, 1997); Hans-Jörg Rheinberger, *Towards a History of Epistemic Things: Synthesizing Proteins in the Test Tube* (Stanford, Stanford University Press, 1997); Karin D. Knorr-Cetina, *Epistemic Cultures: How the Sciences Make Knowledge* (Cambridge, MA, Harvard University Press, 1999).

7 Gilles Deleuze and Félix Guattari, *What is Philosophy*, translated by Hugh Tomlinson and Graham Burchill (London, Verso, 1994), 203.

8 Gilles Deleuze and Félix Guattari, *A Thousand Plateaus: Capitalism and Scizophrenia*, translated by Brian Massumi (London/Minneapolis, University of Minnesota Press, 1987), 409-12.

9 'There's Plenty of Room at the Bottom: An Invitation to Enter a New Field of Physics', *Engineering and Science* 23:5 (1960), 22–36.

10 Drexler, *Engines of Creation*, 3.

11 Identified, for example, by Nikolas Rose in 'The Politics of Life Itself', *Theory, Culture and Society* 18:6 (2001), 1–30.

12 See, for example, Cyrus C. Mody, 'Small, but Determined: Technological Determinism in Nanoscience', *Hyle: International Journal for Philosophy of Chemistry* 10:2 (2004), 99–128.

13 Indeed, Drexler's figure for nanotechnology *par excellence* is the *desktop assembler* in which he suggests it will be possible to fabricate almost anything on a personal computer by simply arranging atoms from the bottom up. See also Gershenfeld, *FAB*.

14 Erwin Schrödinger, *What is Life? The Physical Aspect of the Living Cell* (Cambridge, Cambridge University Press, 1944), 3, 21.

15 See J. Bennett, 'De rerum natura', *Strategies: Journal of Theory, Culture and Politics* 13:1 (May 2000), 9–22.

16 B.C. Crandall, in *Nanotechnology: Molecular Speculations on Global Abundance*, edited by B.C. Crandall (Cambridge, MA, MIT Press, 1996), 5.

17 Jamie Dinkelacker, quoted in Stephen J. Wood., Richard Jones and Alison Geldart, *The Social and Economic Challenges of Nanotechnology* (Economic and Social Research Council, 2003), 26.

18 John von Neumann, 'The General and Logical Theory of Automata', in John von Neumann, *Collected Works* (Oxford, Pergamon Press, 1961), 311.

19 Von Neumann, *Collected Works*, 318.

20 Tomasso Toffoli, 'Occam, Turing, von Neumann, Jaynes: How much can you get for how little? (A conceptual introduction to cellular automata)', *InterJournal Complex Systems* 2 (1994), *http://pm1.bu.edu/~tt/publ/occam.ps*, 1.

21 Toffoli, 'Occam, Turing'.
22 Jean-Marie Lehn, 'Toward Complex Matter: Supramolecular Chemistry and Self-Organisation', *Proceedings of the National Academy of Sciences* 99 (2002), 4763.
23 Ilya Prigogine and Isabelle Stengers, *Order Out of Chaos* (New York, Bantam Books, 1984); Bonta and Protevi, *Deleuze and Geophilosophy: A Guide and a Glossary*; and John Urry, 'The Complexity Turn', *Theory, Culture & Society* 22:5 (October 2005), 1–14.
24 See Brian Wynne, 'Reflexing Complexity: Post-genomic Knowledge and Reductionist Returns in Public Science', *Theory, Culture and Society* 22:5 (October 2005), 67–94. Wynne makes a similar argument in relation to post-genomics and epigenomics.
25 Fritjof Capra, 'Complexity and Life', *Theory, Culture & Society* 22:5 (2005), 33.
26 See Stengers, *Power and Invention: Situating Science.*
27 See Wynne, 'Reflexing Complexity'.
28 The Drexlerian vision of nanotechnology and the possibilities of molecular manufacture — which was later to become highly contested technically — was initially important in the establishment of a National Nanotechnology Initiative (NNI) in the U.S. as a vehicle through which to organize and coordinate research funding. The title of an early NNI brochure, *Nanotechnology: Shaping the World Atom by Atom* (Washington, D.C, NSTC, 1999), indicates the resonances of this vision, which was tacitly accepted by the Clinton administration.
29 Gilles Deleuze, *Difference and Repetition*, translated by Paul Patton (London, Athlone, 1994).
30 Peter Hallward, 'Deleuze and The "World Without Others"', *Philosophy Today* 41:4 (1997), 530–44.
31 See Gilbert Simondon, *L'Individu et sa Genèse Physico-Biologique* (Paris, Presses Universitaires de France, 1964), a section of which is translated as 'The Genesis of the individual' in *Incorporations: Zone 6*, edited by Jonathan Crary and Sanford Kwinter (New York, Zone Books, 1992), 296–319.
32 Deleuze, *Difference and Repetition*, 90.
33 See Martin Heidegger, *The Question Concerning Technology* (New York, Harper & Row, 1977).
34 Leslie White's *The Science of Culture: A Study of Man and Civilization* (New York, Grove Press, 1958) epitomizes the supposed link between technology and human progress in early anthropology and socio-cultural evolutionism. Such thinking also has its contemporary equivalents in texts such as Kurzweil's *The Singularity is Near.*
35 Deleuze, *Negotiations*, 181.
36 Drexler was most thoroughly criticized along these lines by Richard Smalley. See, for example, Richard Smalley, 'Of Chemistry, Love and Nanobots', *Scientific American* (September 2001), 68–9.

37 Smalley, 'Of Chemistry, Love and Nanobots'.

38 Richard A.L. Jones, *Soft Machines: Nanotechnology and Life* (Oxford, Oxford University Press, 2004).

39 See, for example, *Opportunities and Uncertainties* (London, Royal Society and Royal Academy of Engineers, 2004).

40 Bernadette Bensaude-Vincent, 'Two Cultures of Nanotechnology?', *Hyle* 10:2 (2004), 65–82.

41 Jones, *Soft Machines*.

42 *Soft Machines*, 8.

43 *Soft Machines*, 123.

44 Nadrian C. Seeman and Angela M. Belcher, 'Emulating Biology: Building Nanostructures From the Bottom Up', *Proceedings of the National Academy of Sciences* 99 (supplement v2) (2002), 6451.

45 In 'Small, but Determined: Technological Determinism in Nanoscience', Mody suggests that, though nanotechnology has yet to achieve coherence as a field, similar forms of epistemic determinism inform much of the thinking in nanotechnology. He states: 'Nano is still an incoherent mass of often conflicting communities. Determinist arguments advance the particular interests of various kinds of practitioners within this mass, as well as various critics and supporters on the outside' (123).

46 Karen J. Elder, 'Soap and Sand: Construction Tools for Nanotechnology', *Philosophical Transactions of the Royal Society A: Mathematical, Physical and Engineering Sciences* 362:1825 (2004), 2644.

47 Seeman and Belcher, 'Emulating Biology', 6453.

48 See Jones, *Soft Machines*.

49 Compare, for example, Deleuze's sense of the spontaneous internal creative force inherent to technological change with the grand theories of the evolution of technical systems of Bertrand Gille and André Leroi-Gourhan — and by extension Derrida and Steigler. They suggest a grand technological lineage linking contemporary advances with an originary technicity. Though not straying into the positivistic terrain of the social-cultural evolutionists, such a theory irreducibly links technology to the 'human'. Though perhaps respecting the import of the theory of orginary technicity, Deleuze also holds out for a thoroughly non-human, spontaneous evolution in technical systems and objects. See Mackenzie, *Transductions*, in this respect.

50 Michael Conrad, 'Emergent Computation Through Self-Assembly', *Nanobiology* 2:1 (1993), 5–30.

51 Michael Conrad, 'Origin of Life and the Underlying Physics of the Universe', *Biosystems* 42:2–3 (1997), 189.

52 Conrad, 'Origin of Life', 189. Conrad's wider thesis, following Schrödinger, is that self-organizing dynamics of biological systems are an extension of quantum mechanics and general relativity. That is to say that Conrad suggests

a 'great chain of being model' in which life itself develops on the same basic quantum dynamics as the early universe.

53 Susan Oyama, *The Ontogeny of Information: Developmental Systems and Evolution*, second edition (Durham, Duke University Press, 2000), 130.

54 Gilles Deleuze and Claire Parnet, *Dialogues*, translated by Hugh Tomlinson and Barbara Habberjam (New York, Columbia University Press, 1977), 57.

55 Gilles Deleuze, 'Bergson's Conception of Difference', translated by Melissa McMahon, in *The New Bergson,* edited by John Mullarkey (Manchester, New York, Manchester University Press, 1999 [1956]), 42–3.

56 Deleuze and Guattari, *A Thousand Plateaus*, 274.

57 *A Thousand Plateaus*, 353.

Molecular Biology in the Work of Deleuze and Guattari

JOHN MARKS

Abstract
This article looks at Deleuze and Guattari's understanding of molecular biology, focusing particularly on their reading of two highly influential works by the eminent French molecular biologists François Jacob and Jacques Monod, *La logique du vivant* (*The Logic of Living Systems*) and *Le hasard et la nécessité* (*Chance and Necessity*). In these two works, Jacob and Monod present the significance of molecular biology in broadly reductionist terms. What is more, the *lac operon* model of gene regulation that they propose serves to reinforce the so-called Central Dogma of molecular biology, according to which information passes from DNA to RNA to proteins, with no reverse route. However, Deleuze and Guattari discover *intensive* potentials within the descriptions of molecular biology offered by both writers. It is argued that Jacob's work in particular, as it has developed in the years since the publication of *La logique du vivant* in 1970, has itself developed these intensive potentials.

Keywords: molecular biology, Darwinism, reductionism, intensive science

Deleuze, Guattari and biophilosophy

In recent times, there have been several illuminating commentaries on the 'biophilosophy' of Gilles Deleuze.[1] These commentaries have established the importance of developments in the field of biology for Deleuze's thinking in both his own works and those co-authored with Félix Guattari. Key themes that emerge are the influence of Weismann's concept of germ plasm, Darwinism and neo-Darwinism, and Bergson's concept of creative evolution. Taken together, these commentaries also bring out some important tensions. On the one hand, Howard Caygill has portrayed this biophilosophy as shrinking from the rigorously inhuman focus of Darwinian selection in favour of an ethology of the organism's affects. On the other hand, Deleuze and Guattari are seen by Mark Hansen as failing to emphasize sufficiently the affective role of the organism, tending instead to reduce it to the 'epiphenomenon or existential effect' of molecular forces. Also, it seems that Deleuze and Guattari forge a philosophical position that is clearly influenced by neo-Darwinism, but which is also in some senses compatible with complexity theory, and its challenge to the rigours of a Darwinist perspective.[2] It will be argued that these tensions arise from a typically bold, and in many ways prescient, reading of evolutionary theory on Deleuze and Guattari's part. That is to say, they are attracted by the focus upon impersonality in Darwin's and Weismann's thinking, and they incorporate these impersonal mechanisms into their own preoccupation with an immanent plane of *molecular* becoming.

In order to understand Deleuze and Guattari's reading of evolutionary theory, the particular aim here is to explore the role that the scientific field of molecular biology plays in Deleuze's biophilosophy. More precisely, this article will deal with Deleuze's response to the 'molecular' version of neo-Darwinism that was mediated in France through the work of molecular biologists François Jacob and Jacques Monod. Deleuze (and Guattari) read the popularizing texts of both Jacob and Monod, and there are key references to both books in *Anti-Oedipus* and *A Thousand Plateaus*.[3] Ultimately, it will be argued, Deleuze and Guattari's reading of molecular biology is in line with their broader treatment of science. That is to say, they seek to explore the *intensive* potentials within a scientific field that has tended to formulate itself in *extensive*, as well as explicitly reductionist and deterministic terms. So, although their reading of molecular biology is fragmentary and at points highly contentious, it has the virtue of bringing out an intensive dimension that is identifiable in the work of François Jacob, if not Monod.

In order to understand what is meant by 'intensive' here, it is necessary to consider briefly Manuel DeLanda's recent work on the connections between philosophy and intensive science in the work of Deleuze and Guattari.[4] DeLanda focuses on the way in which the extensive structures that we habitually perceive in the world around us emerge from an undifferentiated intensive space (*ISVP*, 27). He explains the relation between *intensive* and *extensive* properties in terms of the difference between *topological* and *metric* spaces (*ISVP*, 45). Topology is defined as the 'least differentiated' form of geometry, in which figures that would be distinct in Euclidean geometry can be 'deformed' into each other. He uses this geometric definition of topology to provide a metaphorical account of the way in which real space is engendered: 'As if the metric space which we inhabit and that physicists study and measure *was born* from a nonmetric, topological continuum as the latter differentiated and acquired structure following a series of symmetry-breaking transitions' (*ISVP*, 26). The dimension of the continuum is central to an understanding of the intensive, along with the notion that an intensive property cannot be divided without producing a 'change in kind'. DeLanda offers as an example the division in temperature created by heating a container from beneath. This temperature division will ultimately give rise to a 'phase transition' and a consequent change in quality of the water (*ISVP*, 27).

As far as DeLanda is concerned, Deleuze's ontology is intensive, in the precise sense that he distinguishes between metric and nonmetric spaces — often formulated in terms of the 'striated' and the 'smooth' — and also in the more general sense that he seeks to avoid the pitfalls of typological and essentialist thinking. This tendency is expressed in Deleuze's concept of the virtual, in that the virtual designates a 'pure multiplicity', rather than the 'possible' (a non-existent identity). Deleuze focuses on the intensive particularity of a given individuation process, such as the way in which an embryo develops into an organism (*ISVP*, 40). Embryology is discussed in relation to the virtual, the realm of Ideas, in *Difference and Repetition*:

> Embryology shows that the division of an egg into parts is secondary in relation to more significant morphogenetic movements: the augmentation of free surfaces, stretching of cellular layers, invagination by folding, regional displacement of groups. A whole kinematics of the egg appears, which implies a dynamic. Moreover, this dynamic expresses something ideal.[5]

As far as biology is concerned, DeLanda associates Deleuze, via his reading of Darwin, with an 'intensive' conception of speciation. That is to say, in what we might call an 'intensification' of Darwin's attack on the notion of fixed species, Deleuze argues that the identity of species is the result of difference. It is for this reason that Deleuze describes Darwinism as a Copernican Revolution.[6] The species does not precede the individual: it is rather individual difference that produces the species. As we will see, Deleuze extends this focus on embryology to encompass a general understanding of the living world not in terms of a hierarchy of distinct ontological levels, but rather as a *flat ontology* of singular, different individuals. 'Population thinking', as this challenge to typological thinking is known, focuses on the differential qualities of all of the features of the organic world. Quite simply, for population thinking, genetic variation is a pre-requisite for the emergence and existence of species (*ISVP*, 48).

Neo-Darwinism and molecular biology

Before looking at Deleuze and Guattari's encounter with Jacob and Monod, it is worth recalling some of the key features of neo-Darwinism, and in particular the importance of molecular biology for this updating of Darwin's theory of evolution. Stuart Kauffman outlines four strands that converge to form what he calls the 'Neo-Darwinian Synthesis'.[7] The first strand is Darwin's theory of evolution. Darwin moves away from the notion of fixed, unchanging species towards the concept of branching phylogenies. Rather than being considered as fixed essences, organisms are now seen in terms of history and contingency, and the image of branching remains at the heart of all thinking about organisms and evolution. The second strand is Mendel's celebrated experimental observations on the laws of heredity, which were rediscovered at the beginning of the twentieth century in the form of 'transmission genetics'. Mendel shows that the genetic material is passed down in the form of 'genes'. The phenotype of the organism is constructed by means of dominant genes inherited from both parents, rather than from a blending of the genetic material. The 'atoms' of heredity pass down unchanged through the generations, and are simply recombined and reshuffled in different ways. Third, is Weismann's doctrine of 'germ plasm'. Weismann formulated his theories several decades in advance of the discovery and analysis of DNA, yet his concept of germ cells containing information that is passed on down the generations anticipates the role that is widely attributed to DNA with the development of molecular biology in the second half of the twentieth century. As Kauffman points out, the notion of germ plasm as a sort of central directing agency has developed into the current concept of the genome controlling the development of the organism: the notion of the genetic 'programme'.[8] The idea that the germ line is shielded from environmental influences also prefigures molecular biology's reinforcement of a Darwinian theory of evolution through adaptation and selection, as opposed to a Lamarckian model of the inheritance of acquired characteristics. The fourth strand of the neo-Darwinian Synthesis is population genetics, which is to say the analysis of the way in which selection makes it possible for the genetic makeup of a population to change over time.

With the development of molecular biology in the post-war period this neo-Darwinian synthesis crystallizes into what Kauffman identifies as a 'core paradigm'. At the heart of this core paradigm is the Darwinian notion that natural selection 'sifts through' the random mutations in the genetic material that appear in the course of reproduction. Within this paradigm, Weismann's notion of germ plasm has been recast as the concept of the genome as a genetic 'programme' that acts as a central directing agency, or 'blueprint' for the development of the organism:

> What complex chemical structures are the carriers of Mendelian genes? How do genes accomplish their transmission? How do such genes become expressed as traits in the offspring? Answering these questions has led to the elucidation of the structure of DNA and the genetic code, the expression of structural genes as specific proteins playing catalytic or structural roles in the developing organism, and the concepts of regulatory genes and 'cybernetic' genetic regulatory circuits governing patterns of gene expression.[9]

This core paradigm is also reductionist in terms of the lines of investigation that it sets up. Reductionism is, of course, the claim that the lowest order components are the key to understanding any entity: the claim that complex things can always be *reduced* to simple, basic building blocks. The central methodological principle is the attempt to break down complex systems into their constituent components. The reductionist tendency of molecular biology as it is formulated from the mid-1950s onwards, is expressed most explicitly in the so-called 'Central Dogma' of molecular biology, originally formulated by Francis Crick. The Central Dogma is based around the scientific claim that the flow of genetic information goes *in one direction* from DNA to RNA to protein. The fact that there is no way for information to travel back into the genetic material effectively rules out the environmental effects on heredity: according to the Central Dogma, there can be no acquired characteristics that are passed on to offspring. In more recent times, Dawkin's concept of the 'selfish gene' has also played a significant role in popularizing the reductionist tendency of much modern biology. According to Dawkins genes are 'selfish', in that DNA is the fundamental unit of inheritance and reproduction — the 'replicator' — and organisms are simply 'vehicles' for the successful transmission of replicators. It should be noted, in passing that, in recent times, the growing field of epigenetics has led to a fairly widespread questioning of the rigidity of the Central Dogma.

Jacob and Monod

Jacob and Monod played a key role in the development of the science of molecular biology. The highpoint of molecular biology as a conceptual science in France came in 1965 when, along with André Lwoff, they were awarded the Nobel Prize for their work on gene regulation in micro-organisms. In his comprehensive survey of molecular biology, Michel Morange goes so far as to claim that the discoveries of Jacob, Monod and Lwoff established the conceptual coherence of molecular biology, since they provided a comprehensive picture of the way in which the genetic material manages information.[10] In simple terms, the so-called 'operon' model showed how certain 'regulatory' genes control the activity of 'structural' genes that code for enzymes or proteins. The distinction between regulatory genes and structural genes was also significant in philosophical terms, in that it added a crucial dimension to the informational understanding of life that had begun to develop in the 1940s. In identifying two types of gene, Jacob and Monod were also effectively identifying two types of information at work in the construction of living organisms. More generally, the way in which they conceptualized the significance of their work on the lactose operon is at the root of the widespread notion that DNA is, as Kauffman puts it, 'a kind of biochemical computer which executes a developmental program leading to the unfolding of ontogeny'.[11]

Both Jacob and Monod engaged in the task of presenting what they saw as the significance of the 'new' science of molecular biology to a public readership. Monod's *Chance and Necessity* and Jacob's *The Logic of Living Systems*, both originally published in 1970 in France, stand as definitive statements of the molecular reinforcement of the neo-Darwinist synthesis. Monod's book, published in 1970, attempted to explain to a general audience how recent advances in genetics challenged what he called the 'anthropocentric illusion'.[12] Writing in the wake of the so-called Lysenko affair, Monod rejects what he sees as a series of 'anthropocentric' illusions. He argues that the 'ancient covenant' between man and nature has been definitively broken, and that it is no longer possible for 'man' to project human values into the natural world. Modern societies enjoy the benefits of science whilst clinging to anachronistic values. For Monod, the Western blend of Judeo-Christian religiosity, scientistic progressivism and utilitarian pragmatism constitutes an 'animist' illusion, as does the Marxist commitment to dialectical materialism. Animism, as Monod defines it, consists of man's projection into inanimate nature of his awareness of human 'teleonomy' (the fact that organisms are endowed with a purpose that is inherent in their structure and which determines their behaviour). There is, he argues, no universal theory that can encompass the biosphere; chance alone is the source of innovation and creation. Molecular biology not only reveals the fact that the living world runs with mechanical precision, but also confronts us with a world that is entirely indifferent to human values and preoccupations. Ultimately, Monod's avowedly philosophical position is a curious mixture of Cartesian mechanicism and the existentialism of Albert Camus. His scientific stance is rigidly reductionist, in that DNA is portrayed as a universal machine that governs the production of the whole variety of living organisms.

Jacob's *The Logic of Living Systems* is also based on a reductionist conception of molecular biology, and this reductionist perspective is expressed in terms of metaphors drawn from linguistics. The DNA in the cell, Jacob claims, constitutes a 'dictionary' comprising sixty-four genetic terms, and each of the twenty amino acids in the cell corresponds to a series of 'synonyms' amongst these terms (*LLS*, 276). Jacob goes on to claim in broader evolutionary terms that molecular biology provides two major insights into the nature of living systems. First, the organism is nothing more than the realization of a programme prescribed by heredity. Second, evolution is not a teleological process governed by some higher power, but rather an impersonal process of selection between genetic programmes. The genetic programme cannot 'learn' from experience, but the genetic message that is passed from generation to generation is able to integrate all the results of past reproductions. It is, Jacob claims, like a 'text without an author' (*LLS*, 287).

In a general sense, Jacob's philosophical approach is more subtle and less ideologically influenced than Monod's. Jacob is clearly influenced by structuralist interpretations of history, for example, in the way that he traces a series of epistemic shifts in the history of biology. First, in the seventeenth century, the notion of life as a *visible structure* replaces the view of life as a form of sacred text. Then, at the end of the eighteenth century life begins to be understood in terms of *organization*. At the start of the twentieth century the model of *chromosomes* and *genes* replaces the episteme of organization. Finally, in the second half of the twentieth century, the *molecule* emerges as the essential unit of living systems. Biology also produces, according to Jacob, one major theory, as opposed to the numerous generalizations characteristic of the history of biology, in the shape of Darwin's theory of evolution. The theory of evolution has the dual effect of making biology a discipline concerned with the processes of time and history in the natural world, as well making biologists like Jacob reflect on biology as a historically situated discipline. That is to say, in philosophical terms, the theory of evolution inserts a concept of *difference* into biology. Small differences produced by minor glitches — 'copying errors' or 'recombinatorial spoonerisms' as Jacob puts it — lead to phenotypic differences in the population upon which pressures of selection come to bear (*LLS,* 292). The individual that carries this new genetic information is subsequently 'put to the test' of reproduction and, if it is successful, it plays its part in favouring the propagation of the particular programme that has created it. By means of this process, the genetic text is effectively modified and moulded by environment, but only in a 'roundabout way', Jacob emphasizes, by means of differential success in reproduction.

In this way, Jacob, along with Monod, sets out a version of Darwinian evolutionary theory that is apparently bolstered by the reductionist insights of molecular biology. The genetic 'text' functions as a motor for evolution by virtue of the fact that it is on one level copied with remarkably accuracy, whilst displaying crucial minor variations over time. Monod, despite his occasionally idiosyncratic philosophical stance, captures this elegantly in the claim that 'evolution is chance caught on the wing': the machine-like replication of DNA is highly efficient at capturing and in turn replicating minor differences.

The 'rhetorics' of the operon model

One of the most intriguing aspects of Jacob and Monod's work is the way in which they both enthusiastically engaged in establishing what might be called a 'rhetorics' of molecular biology. Monod's claim that evolution is chance 'caught on the wing' and Jacob's later claim that 'evolution tinkers' stand as key metaphorical formulations — along with Dawkin's 'selfish gene' — of the neo-Darwinian synthesis that was crystallized by molecular biology in the 1950s and 1960s. Richard Doyle, in his broadly deconstructionist *On Beyond Living,*[13] perceives this rhetoric at work not only in Jacob and Monod's 'philosophical' work, but also in the apparently dry scientific language of the paper in which they made public their work on the lac operon model: 'Genetic Regulatory Mechanisms in the Synthesis of Proteins'.[14] Doyle sees the operon model as a rhetorical strategy as much as an influential scientific discovery.

In scientific terms, the details of the operon model are relatively straightforward. Jacob and Monod identified 'repressor' and 'promoter' genes that switch other genes on or off. Consequently, the operon model of enzyme induction hypothesizes that, when lactose is absent, the repressor molecule attaches itself to the operator gene, and so switches it off. When lactose is present, on the other hand, the repressor molecule cannot attach itself to the operator gene, so it remains switched on. This means that the structural gene is operational, and codes for beta-galactosidase, which breaks down lactose. It seemed that Jacob and Monod had located a simple mechanism that would help to explain the self-evident fact that genes cannot build an organism by acting in the same way at all times. However, as Doyle presents it, Jacob and Monod present the significance of their scientific work on enzyme induction in such a way as to police 'what we might hazard to call the literal boundary between the inside of the organism and the outside, and the inside and outside of science'.[15] Jacob and Monod were not simply transmitting the technical details of an important scientific discovery: they were also reinforcing a largely reductionist and somewhat mechanistic view of the world. As Doyle shows, the phenomenon of enzyme *adaptation*, whereby enzymes are produced in cells in response to environmental agents — such as lactose in the case of *E. coli* — constituted a challenge to the notion that DNA acts as the site of control, the central and unique programming centre for the synthesis of proteins. In redefining the phenomenon as a process of *induction*, and thus effectively reinscribing environmental effects within the all-powerful domain of DNA, Jacob and Monod were able to fight off this challenge. In short, the rhetorical effect of the operon model is that DNA retains its role as the 'Master Molecule', as Doyle puts it, and the significance of 'external' influences such as the growth and development of the organism within a particular environment, are downplayed, if not discounted. In rhetorical terms, the operon model operates a temporal and material sleight of hand, which is achieved by the occlusion of the organism:

> The claim that the genome, if not the structural genes alone, is necessary and sufficient for the 'definition' of the structure of proteins depends upon the erasure of the nucleic acids' dependence on the cytoplasm and the organism, an erasure that is made possible by the slippage between 'instruction' and 'construction' in Jacob and Monod's rhetoric of definition. This slippage effaces the process of protein synthesis, a process that requires time and an organism, not just a genetic fiat.[16]

In this way, Jacob and Monod attribute to DNA a position of 'impossible retroactivity', as Doyle puts it: DNA defines the organism before it exists, and yet it depends on the cellular machinery of the organism in order to express itself.

Jacob: population-thinking and embryology

What, then, do Deleuze and Guattari find of use in the scientific work of Jacob and Monod, given that the latter are clearly reductionist neo-Darwinists? Todd May has recently pointed to the fact that Deleuze and Guattari bring out an intensive potential in Monod's *Chance and Necessity* by focusing on Monod's analysis of allosteric enzymes.[17] Monod indicates that these enzymes have self-ordering, 'cognitive' properties, and his analysis focuses on the way in which life emerges in a 'random' manner from the chance coming together of enzymes in a favourable environment. As May points out, Deleuze and Guattari interpret the significance of allosteric proteins in a slightly different way:

> Allosteric and other enzymes contain the capacity for all sorts of combinations at the preindividual level. Moreover, these combinations, because they are the product of chance, might have been and might have become otherwise. The molecular level is a virtual realm of intensities, a field of difference that actualizes itself into specific biological arrangements.[18]

However, as already suggested, Jacob's work proves to be even more promising in terms of its intensive potentials. For one thing, Jacob has continued to publish 'popular' science work up until the present day, enabling him to assimilate and respond to challenges to the paradigm of molecular biology that he set out with Monod, who died in 1976.[19] Also, Jacob is aware of issues that relate to the 'rhetorics' of science. So, the closing lines of *The Logic of Living Systems* anticipate as yet undiscovered forms of organization, further 'Russian dolls' that as yet are undetected: 'Today the world is messages, codes and information. Tomorrow what analysis will break down our objects to reconstitute them in a new space? What new Russian doll will emerge?' (*LLS*, 324). In fact, just as the operon model constitutes the 'first wrinkle' on the face of the central dogma, so the closing paragraphs of *The Logic of Living Systems* point to the fact that reductionism cannot fully explain the behaviour of matter in the living world. With the benefit of hindsight, these closing lines seem to point towards the concept of 'emergence' — the phenomenon of 'integration' reveals a world of stratified levels in which the properties of each level cannot be predicted from the components of the lower level:

> Integration changes the quality of things. For an organization often possesses properties that do not exist at the level below. These properties can be *explained* by the properties of the components; they cannot be *deduced* from them. This means that a particular integron has only a certain probability of appearing. All forecasts about its existence can only be statistical. This applies equally to the formation of beings and things; to the constitution of a cell, an organism, or a population, as well as of a molecule, a stone or a storm. (*LLS*, 323)

Seen in this light, *The Logic of Living Systems*, as well as constituting a classic statement of molecular biology as an essentially reductionist scientific field, can also be seen as anticipating the intensive reading of Darwin that DeLanda draws out in the work of Deleuze. Again, it is Darwin's theory of evolution that provides a route into the realm of the intensive. So, in an analysis that is clearly influenced by Foucault's work on epistemology, Jacob claims that, until the middle of the nineteenth century, the living world was seen as a continuous series of forms. The hierarchy of living forms was continuous — breaks were simply the result of faulty observation — and was proof of a preconceived harmony. Evolution, of course, attacked the very foundations of this order: living beings were henceforth part of a system that regulated itself from within. Evolution was one expression of a wider epistemic shift in the scientific understanding of objects. Previously, objects, including living beings, had essentially been the reflection of a 'type', and any deviations from the type were seen as negligible defects. Now, however, it is, on the contrary, the singular nature of each individual that is important:

> Each member of the group is unique. There is no longer a pattern to which all individuals conform, but a composite picture, which merely summarizes the average of each individual's properties. What has to be known, then, is the population and its distribution as a whole. The average type is just an abstraction. Only individuals, with their particularities, differences and variations, have reality. (*LLS*, 173)

In short, Jacob clearly formulates a form of population thinking that conceptualizes difference and heterogeneity as intensive properties. A similar form of population thinking, which acknowledges a debt to Darwin, is outlined in *A Thousand Plateaus*, where Deleuze and Guattari emphasize the 'nomadic' dimensions of Darwinism, insofar as it moves away from essentialist and typological thinking by substituting populations for types and differential relations for degrees of perfection. Deleuze and Guattari, in turn, identify an intensive potential in what they call, drawing on Monod, 'molecular Darwinism'. As far as they are concerned, molecular biology confirms this Darwinian conception of population by showing that the individual is always caught up in what they call 'molecular populations and microbiological rates' (*ATP*, 49).

The other main area in which Jacob moves in the direction of an intensive understanding of biology is in his recognition, particularly in the 1980s, that embryology poses questions that molecular biology, with its concentration on linear, extensive conceptualization of the genetic code, finds it difficult to answer. For example, in *The Possible and the Actual*, Jacob admits that molecular biology can describe in detail the molecular description of a mouse, as well as being able to show how it breathes and digests food. However, it cannot explain how the mouse is formed from a single egg cell. The notion that a 'Laplacian demon' would be able to analyse the molecular structures of the fertilized egg and thus describe the future organism remains entirely hypothetical, since molecular biology does not yet have the algorithm to understand just how the genetic programme plays itself out. At this point, Jacob makes a striking admission:

> For the only logic that biologists really master is one-dimensional. As soon as a second dimension is added, not to mention a third one, biologists are no longer at ease. If molecular biology was able to develop so rapidly, this is largely because, in biology, information happens to be determined by simple linear sequences of building blocks, bases in nucleic acids and amino-acids in protein chains. Thus the genetic message, the relations between the primary structures, the logic of heredity, everything turned out to be one-dimensional.[20]

The development of the embryo cannot be accounted for by this linear model of molecular biology. Biology cannot explain the processes by which the linear, one-dimensional sequence of the bases in the genetic material is translated into two-dimensional cell layers, three-dimensional tissues and organs, and 'four-dimensional' behaviours. As we have seen already, embryogenesis plays a key role in Deleuze's biophilosophy. In *Difference and Repetition* the transformations undergone by the embryo are highlighted as being definitively *intensive*.[21] The embryo, as a 'larval subject', undergoes a 'pure spatio-temporal dynamism', an intense process of twisting and folding that would literally break the skeleton of an adult organism.

Creative involution

Jacob's interpretation of molecular biology also converges with Deleuze and Guattari's more general conviction, which is most comprehensively formulated in *A Thousand Plateaus*, that extensive organic forms in fact constitute a single organic stratum, a continuum. The notion of an intensive stratum that provides an underlying unity to the living world is expressed as a preference for Étienne Geoffroy Saint-Hilaire in *A Thousand Plateaus* (*ATP*, 45-7). Geoffroy asserted that there is no particular form of matter that is particular to the organic stratum, but at the same time he claimed that this stratum is characterized by a unity of composition; that the same components are reconfigured in different ways across the stratum. It is clear that Deleuze and Guattari regard Geoffroy as anticipating the insights of molecular biology:

> It is of little or no importance that Geoffroy chose anatomical elements as the substantial units rather than protein and nucleic acid radicals. At any rate, he already invoked a whole interplay of molecules. The important thing is the principle of the simultaneous unity and variety of the stratum: isomorphism of forms but no correspondence; identity of elements or components but no identity of compound substances. (*ATP*, 46)

Over the course of his writing on biology, Jacob develops a strikingly similar view that molecular biology has revealed that basic genetic structures are shared across the living world. It as if there is a base stratum of genetic material that is held in common by all organisms. In this sense, the living world is, Jacob claims, like a giant children's construction set, as if all living beings are constructed from different combinations of a relatively restricted set of components: 'The whole of the living world looks like some kind of giant Erector set. Pieces can be taken apart and put together again in different ways, to produce different forms. But fundamentally the same pieces are always retained.'[22] That is to say, evolution works in a sort of *ad hoc* manner, putting together different combinations of the limited components that come to hand.

In light of this 'flat ontology', Deleuze and Guattari push their intensive reading of evolutionary theory in *A Thousand Plateaus* in a direction that goes so far as to challenge the precepts of molecular Darwinism. They invoke the term 'involution' in order to distinguish this formulation of 'becoming' from the emphasis placed by the theory of evolution on filiation passing genetic material down through successive generations. By focusing on molecular processes of individuation that precede differences mediated at the molar level by individuals, Deleuze and Guattari claim that, at the level of the population, the genetic 'code' is always subject to 'decoding'. In support of their argument, they refer here to Jacob's description of genetic mutations in *The Logic of Living Systems*, where he sets out two different kinds of event that increase the amount of genetic information in bacterial cells (*LLS,* 290-2). The first type of event would be a segment being copied twice when a chromosome is being reproduced. Deleuze and Guattari pick up on Jacob's assertion that the second copy is then able to mutate freely over time. The second event that Jacob identifies in bacteria involves the occasional transfer of genetic material from one cell to another. What Deleuze and Guattari call 'transversal communications' between heterogeneous populations constitute a challenge to evolution's preoccupation with filiation:

> In addition, fragments of code may be transferred from the cells of one species to those of another, man and Mouse, Monkey and Cat, by viruses or through other procedures. This involves not translation between codes (viruses are not translators) but a singular phenomenon we call surplus value of code, or side-communication. (*ATP*, 53)

Against the rigid determinism of neo-Darwinism, Deleuze and Guattari also suggest that *environment* has some agency in the determination of the code, and also that 'side communication' takes place, synchronically as it were, across species. The overall effect in *A Thousand Plateaus* is to locate genetic change in a shifting, fluid set of milieus that effectively compose the individual organism. In short, Deleuze and Guattari appropriate the *molecular* dimension of neo-Darwinism, but largely reject its genetic determinism. Of course, it may be argued that, in placing such emphasis on transgenic transfers, Deleuze and Guattari misrepresent the central insights of molecular Darwinism. However, even if this is the case, the reading of molecular biology that Deleuze produces — in his own work, as well as with Guattari — does have the virtue of drawing out intensive potentials.

Conclusion: the double helix as 'superfold'

One of the rare occasions when Deleuze talks directly about molecular biology comes in 'On the Death of Man and Superman', the appendix to his book on Foucault, published in French in 1986.[23] Here, Deleuze returns to Foucault's controversial speculations on the 'death of man' in the 1960s. Deleuze's purpose is to re-state the importance of Foucault's formulation, by emphasizing that it is a question of the composition of forces and forms. The form of 'Man' contains and constrains life, and the Nietzschean figure of the 'superman' holds out the promise of freeing this life, to the benefit of another form that might supersede 'Man'. However, although Deleuze wishes to defend the importance of Foucault's claim, he also finds it 'peculiar' that Foucault should privilege language as the particular form — as opposed to life and labour — that regroups itself as a form external to Man. It seems rather, to Deleuze, that labour and life have been able to regroup themselves as what we might think of as 'inhuman' forms, in the same way that language broke free from nineteenth-century humanist linguistics to regroup as an autonomous, self-referential 'being of language'. The key developments that have led to a parallel regrouping of life and labour are molecular biology and information technology. Molecular biology enabled life to be understood in terms of the genetic code, and labour to be theorized in terms of cybernetics and information technology.[24] The discovery of the material structure of DNA means that 'Man' can no longer be conceived of according to the structural rules of what Foucault calls the modern episteme; as both subject and object of his own understanding.[25] Instead, 'man' becomes an 'afterman' [*surhomme*], and finitude gives way to a play of forces and forms that Deleuze calls 'unlimited-finite' [*fini-illimité*]. Quite simply, an infinity of beings can apparently be constituted from the finite number of four bases from which DNA is constructed:

> It would be neither the fold nor the unfold that would constitute the active mechanism, but something like the *Superfold*, as borne out by the foldings proper to the chains of the genetic code, and the potential of silicon in third-generation machines, as well as the contours of the sentence in modern literature, when literature 'merely turns back on itself in an endless reflexivity'.[26]

Viewed through the lens of DeLanda's recent reading of Deleuze in terms of intensive science, the DNA 'superfold' that Deleuze identifies here can be regarded as a topological, intensive space. To illustrate this shift, Deleuze alludes to Rimbaud's enigmatic claim that the man of the future will be 'filled' [*chargé*] with animals. In Deleuze hands, Rimbaud's phrase elegantly evokes the epistemological challenges currently posed to us in a post-genomic world in which we are increasingly forced to view life in the context of a 'trans-human', intensive genetic continuum.

NOTES

1 See Mark Hansen, 'Becoming as Creative Involution?: Contextualizing Deleuze and Guattari's Biophilosophy', *Postmodern Culture* 11:1 (September 2000), http://www3.iath.virginia.edu/pmc/text-only/issue.900/11.1hansen.txt; Keith Ansell Pearson, *Germinal Life: The Difference and Repetition of Deleuze and Guattari* (London, Routledge, 1999); Howard Caygill, 'The Topology of Selection: The Limits of Deleuze's Biophilosophy', in *Deleuze and Philosophy: The Difference Engineer*, edited by Keith Ansell Pearson (London, Routledge, 1997), 149–62.

2 Ansell Pearson points in particular to the influence of complexity theory on *A Thousand Plateaus*, in the sense that Deleuze and Guattari repeatedly emphasize the 'complex' interaction of organism and environment. Along similar lines, Hansen suggests that *A Thousand Plateaus* might usefully be recast in the light of complexity theory's focus on self-organization, a perspective that is implicit but undeveloped throughout the book.

3 Gilles Deleuze and Félix Guattari, *Anti-Œdipus: Capitalism and Schizophrenia*, translated by Robert Hurley, Mark Seem and Helen R. Lane (London, Athlone, 1984); *A Thousand Plateaus: Capitalism and Schizophrenia*, translated by Brian Massumi (London/Minneapolis, University of Minnesota Press, 1987). Henceforth *ATP*.

4 Manuel DeLanda, *Intensive Science and Virtual Philosophy* (London, Continuum, 2002). Henceforth *ISVP*.

5 Gilles Deleuze, *Difference and Repetition*, translated by Paul Patton (London, Athlone, 1994), 214.

6 *Difference and Repetition*, 249.

7 Stuart A. Kauffman, *The Origins of Order: Self-Organisation and Selection in Evolution* (Oxford, Oxford University Press, 1993), 5–15.

8 Kauffman, *The Origins of Order*, 8.

9 *The Origins of Order*, 8.

10 Michel Morange, *A History of Molecular Biology*, translated by Matthew Cobb (Cambridge, Mass./London, Harvard University Press, 1998). See particularly Chapter 14.

11 Kauffman, *The Origins of* Order, 411.

12 Jacques Monod, *Chance and Necessity: An Essay on the Natural Philosophy of Modern Biology*, translated by Austryn Wainhouse (London, Collins, 1972); François Jacob, *The Logic of Living Systems: A History of Heredity*, translated by Betty E. Spillman (London, Allen Lane, 1974), henceforth *LLS*.

13 Richard Doyle, *On Beyond Living: Rhetorical Transformations in the Life Sciences* (Stanford, Stanford University Press, 1997).

14 François Jacob and Jacques Monod, 'Genetic Regulatory Mechanisms in the Synthesis of Proteins', *Journal of Molecular Biology* 3 (1961), 318–56.

15 Doyle, *On Beyond Living*, 66.

16 *On Beyond Living*, 75.
17 Todd May, *Gilles Deleuze: An Introduction* (Cambridge, Cambridge University Press), 90–2.
18 *Gilles Deleuze: An Introduction*, 91.
19 François Jacob, *The Possible and the Actual* (New York, Pantheon Books, 1982); *Of Flies, Mice and Men*, translated by Giselle Weiss (Cambridge, Mass./London, Harvard University Press, 1999).
20 Jacob, *The Possible and the Actual*, 44.
21 Deleuze, *Difference and Repetition*, 118.
22 Jacob, *Of Flies, Mice and Men*, 80.
23 Gilles Deleuze, *Foucault*, translated by Paul Bové (Minneapolis/London, University of Minnesota Press, 1986).
24 Deleuze, *Foucault*, 131.
25 For a full discussion of Deleuze's reading of Foucault and its relevance to the Human Genome Initiative see Paul Rabinow, *Essays on the Anthropology of Reason* (Princeton, New Jersey, Princeton University Press, 1996), 91–111.
26 Deleuze, *Foucault*, 131.

Science and Dialectics in the Philosophies of Deleuze, Bachelard and DeLanda

JAMES WILLIAMS

Abstract
This article charts differences between Gilles Deleuze's and Gaston Bachelard's philosophies of science in order to reflect on different readings of the role of science in Deleuze's philosophy, in particular in relation to Manuel DeLanda's interpretation of Deleuze's work. The questions considered are: Why do Gilles Deleuze and Gaston Bachelard develop radically different philosophical dialectics in relation to science? What is the significance of this difference for current approaches to Deleuze and science, most notably as developed by Manuel DeLanda? It is argued that, despite its great explanatory power, DeLanda's association of Deleuze with a particular set of contemporary scientific theories does not allow for the ontological openness and for the metaphysical sources of Deleuze's work. The argument turns on whether terms such as 'intensity' can be given predominantly scientific definitions or whether metaphysical definitions are more consistent with a sceptical relation of philosophy to contemporary science.

Keywords: science, dialectics, continuity, discontinuity, explanation, progress, critique, transcendental philosophy

Introduction

Though Bachelard and Deleuze write on science and philosophy at different ends of the twentieth century and in very different ways (Bachelard mainly in the first half and mainly as philosophy of science; Deleuze in the latter half and as part of a wider metaphysics), they both respond to the problem of how to learn from revolutions in science and how to learn from the views of reality provided by scientific theories.[1] Their responses are three-fold: First, there is the question of how new scientific models should influence philosophy, that is, how philosophical concepts and methods should change in view of new discoveries. Secondly, there is the question of the philosophical significance of the possibility of scientific revolutions, that is, how philosophy should take account of this possibility (for example, in terms of the role of error and the limitations of scientific knowledge). Thirdly, there is the question of the implications for the relation of philosophy to science: given the possibility of scientific revolution, which one should claim precedence over the other in determining knowledge of reality, or are other kinds of relation more appropriate?

Manuel DeLanda's reading of Deleuze is strong and influential with respect to the first question. He explains the influence of new sciences and scientific discoveries on Deleuze's work and the parallels between Deleuze's philosophy and, for example, complexity theory, in order to defend the claim that Deleuze is opposed to earlier scientific theories and capable of learning lessons from new discoveries better than other thinkers.[2] But these strengths run the risk of drawing attention away from the second and third questions, in particular, by over-stressing the link between Deleuze and a particular science. This essay returns to the earlier contrast between Deleuze and Bachelard, where these latter questions are taken much further, in order to raise a series of possible objections to the DeLanda position — without wishing to dissent from his important conclusions about the importance of particular sciences for illustrating, understanding and applying Deleuze's philosophy.

The following passage from the introduction to DeLanda's *Intensive Science and Virtual Philosophy* shows the kind of difficulty I am interested in:

> Deleuze replaces the false genesis implied by these pre-existing forms which remain the *same* for all time, with a theory of morphogenesis based on the notion of the different. He conceives of difference not negatively, as lack of resemblance, but positively or productively, as that which drives a dynamical process. The best examples are intensive differences, the differences in temperature, pressure, speed, chemical concentration, which are key to the scientific explanation of the genesis of the form of inorganic crystals, or of the forms of organic plants and animals.[3]

The problem lies with the basis for the claim to falsehood in the first sentence, allied to the points about theory and explanation in the same sentence and the final one. To what degree is DeLanda's point about false philosophical claims dependent on scientific theories, rather than on wider philosophical arguments? Is Deleuze's philosophy identified with scientific theories and explanations, as often seems to be the case in DeLanda's book, or is there an independence of the two, as indicated in his use of the concept 'example', in the above passage? It is the idea that *Deleuze* replaces the false genesis that I find most troubling, since it is either that case that this idea of falsehood is philosophical, in which case it clashes with Deleuze's critique of the notion of falsehood as central to philosophical thought, or it is the case that the idea of falsehood is scientific, but then Deleuze has no business incorporating it uncritically into philosophical concept construction, unless he is taking a hard rationalist turn — something that would contradict his novel version of Hume's empiricism, as developed first in his earliest work *Empiricism and Subjectivity*.[4]

In terms of these questions about error, Deleuze and Bachelard's positions allow for different versions of philosophical dialectics that relate philosophy to science; these versions can be separated through a distinction drawn between different philosophical and scientific notions of falsehood and method. However, for both philosophers, scientific method is supplemented by philosophical methods, dialectics and deductions; these retain a critical stance with respect to scientific method and theories. This stance allows for changes in scientific theories, but also — and more importantly — for scepticism with respect to the latest science and an awareness of the implications of earlier failures. This distance seems to be lacking in DeLanda's account, for example, where he fuses philosophical and scientific explanation in his account of the philosophical concept of multiplicity through a scientific explanation of morphogenesis: 'To anticipate the conclusion I will reach after a long and technical definitional journey: multiplicities specify the structure of spaces of possibilities, spaces which, in turn, explain the regularities exhibited by morphogenetic processes' (*ISVP*, 10).

The divergence between Deleuze and Bachelard takes place around the concepts of completeness and continuity. Different views of both concepts lead to opposed definitions of synthesis in relation to the new in dialectics. The source of this difference lies in the role of scientific method in guiding the definition of philosophical dialectics. Bachelard constructs a dialectics around the problem of how to think methodologically given a demand for completeness but a lack of continuity in scientific revolutions. He is responding to the challenge of adding the new to that which is already relatively known, whilst accepting that there can be no final perfect fit between them. Science shows reality to be discontinuous, but thought can bridge this discontinuity through the synthetic transformations and implicit relativity of dialectics. The status of propositions changes through successive adjunctions. There are no ultimate truths that are not open to revision after experimentation.

Deleuze constructs his dialectics around the problem of how to affirm a productive real continuity through a search for completeness, whilst also responding to the proposition that continuity is never a matter of identities or representations. In other words, we can never represent or identify continuity, even relatively and in an open-ended transforming way. Yet reality is continuous and it is possible to speak of better or worse affirmations of that continuity in accordance with individual problems. This is the paradoxical challenge of his dialectics. It is important to note that Deleuze's arguments for continuity are not based on science, but on transcendental deductions for given experiences, as shown in science but also in other forms.

DeLanda's account of Deleuze's philosophy contrasts with both of these positions by mapping a scientific account of continuity and of discontinuity onto a philosophical account, such that no significant distance lies between them. This can be shown, for example, in his shift from the notion that the scientific account provides a metaphor for philosophy to the view that it provides concrete material for the construction of the philosophical model: 'To get rid of the metaphorical content and to show in what sense the series extending from singularities are nonmetric (thus capable of forming a virtual continuum) I will need to introduce one more technical term, that of an infinite ordinal series' (*ISVP*, 73). Though this is a highly effective way of tracing some of the sources of Deleuze's work and of explaining it, it also runs the risk of identifying scientific theory with philosophical theory such that should the former be shown to be false, then the latter will also fail. It is exactly this problem that Deleuze and Bachelard respond to by setting accounts of continuity and discontinuity into a dialectics where philosophical accounts of reality incorporate lessons leant from the fallibility of scientific theories. However, these responses are also in opposition with one another, since Bachelard emphasizes philosophical discontinuity as a parallel to discontinuities in science, whereas Deleuze emphasizes continuity as a condition for discontinuity. To sum up, Deleuze and Bachelard place the sceptical and questioning role of philosophy at different methodological points, whereas DeLanda appears to underplay this role by subsuming it into a distinction between true and false scientific models.

Dialectics and explanation

In his introduction to his *Le nouvel esprit scientifique*[5] Bachelard sets out a philosophical programme responding to the scientific discoveries of the late nineteenth and early twentieth centuries. His main argument is that a new scientific spirit demands changes in philosophical views of reality and method: 'Sooner or later, scientific thought will become the fundamental theme of philosophical polemics; this thought will lead intuitive and immediate metaphysics to be substituted by objectively rectified discursive metaphysics' (*NES*, 6). Philosophy will become a dialectical method that privileges objective scientific discoveries. However, it is very important to realize that, by objective rectification, Bachelard does not mean a direct response or reflection on pure facts or data. On the contrary, objective rectification is already a scientific dialectics, where theories are always part of an ongoing debate around new discoveries, to the point where there are no pure objective facts: 'If immediate reality is only a simple pretext of scientific thought and no longer an object of knowledge, we shall have to pass from the *how* of description to theoretical *commentary*' (10). Thus truth becomes not a matter of objectivity, but of debate: 'Every new truth is born despite the evidence, every new experience is born despite immediate experience' (11).

Modern science demands a dialectical position between theory and fact, indeed, between theories, facts and further theories. It is not that facts contradict theories, neither is it that theories straightforwardly contradict one another. Rather, the nature of the debate is a two-directional one. Facts only appear thanks to theories, notably, on how to simplify reality so that it may reveal facts. Theories make claims about reality that are undermined by the complexity revealed by scientific discovery. There is a dialectics between the need to simplify and the discovery of complexity. This is not a nihilistic relativism. Bachelard gives precise descriptions of this dialectics and of a relativism associated with progress and projects, rather than with the danger of some generalized doubt: 'In fact, as soon as the object is presented as a complex of relations, it must be apprehended through multiple methods. Objectivity cannot be separated from the social character of the task. We can only arrive at objectivity by exposing a method of objectification in a discursive and detailed way' (*NES*, 16). He thinks that this detailed and precise form of dialectics can only come from science (18). Moreover, it is as a model of a psychological process that science stands out. This psychology is not one of negation or opposition, but one of synthesis in terms of rectification and precision: 'an empirical rectification is joined to a theoretical precision' (19). Modern science teaches us that theories are rectified and added to through new empirical discoveries. The techniques surrounding experiments and the things tested-for are made more precise in the light of theory.

Psychologically, modern science does not evolve through the discarding of theories through counterfactuals or simple falsifications, rather, theories have to be included in greater and sometimes looser collections of theories. We should seek completeness through the addition of complementary theories. Complementary does not mean fully unified. Neither does it mean contradictory. These aspects become clearer in the final chapter of *Le nouvel esprit scientifique* on non-Cartesian epistemology. First, like Deleuze's work on biology and thermodynamics in *Difference and Repetition*, Bachelard is concerned with the way modern science complicates our experience of the world, rather than explaining it.[6] Where Descartes seeks simplifications with great explanatory power, Bachelard sees the emergence of the idea of an 'essential complexity of elementary phenomena' (*NES*, 143). These cannot be reduced legitimately and methodologies that do so must be mistaken.

Deleuze tempers this positive side with the view that all sciences must necessarily also have an explanatory side. So, in contrast to Bachelard, Deleuze refuses the radical view of modern sciences as going beyond explanation in the dialectical privileging of complication through theory. Instead, the desire for completeness is already a sign of an explanatory simplification of the richness of a prior intensity. This distinction between the critical role of intensity and an explanatory role is very important in understanding some of the possible limitations of DeLanda's approach, since his interpretation of intensity as explanatory (albeit with important qualifications regarding the impossibility of identifying intensity) runs the risk of fixing intensity in terms of given explanations and the growing body of explained material. In other words, instead of providing the basis for a constantly renewed critical function, intensity becomes part of a vast and gradually filled-in account of reality. This account is open to failure due to new scientific discoveries and to ongoing philosophical scepticism on all sorts of possible bases, such as moral objections or critical distinctions between different fields.

It is not that Bachelard ignores the pedagogical side of modern science — quite the contrary. It is that he situates pedagogy after discovery, as if the simplification demanded by it remains at arms length from discovery proper. Deleuze sees explanation as a necessary betrayal of complexity even in the most pure moments of scientific advance. The problem of explanation is not pedagogical, but proper to scientific method. The question is whether DeLanda underplays explanation as necessarily problematic, for example, by identifying intensity as a philosophical concept with intensive processes in models for scientific explanation: 'The intensive processes that create these materials are another example of a process of progressive differentiation, one which starts with a population of relatively undifferentiated cells and yields a structure characterized by qualitatively distinct cell types' (*ISVP*, 54). The difficulty with this kind of identification comes from the different criteria at work in scientific explanation (for example, where an explanation brings a case under a given law) and philosophical criteria (for example, in Deleuze's controversial use of interest and the creation of truth as indicators of philosophical worth).

Bachelard retains a distance from specific scientific advances and theories by defining synthesis in modern science in terms of a priori mathematical syntheses: not as a move to find objective unity, but as a way of bringing together an objective plurality through the way mathematics brings complementary fields together without reducing them to one another (his main examples for this are the relation between Euclidean and Non-Euclidean mathematics and their relation to Newtonian physics and the physics of relativity). Mathematical synthesis sheds results by adding fields, but does not depend on presumptions of a prior identity: 'That mathematical description is not clear through its elements. It is only clear in its achievement through a sort of consciousness of its synthetic value (...) all basic notions can in some way be doubled; they can be bordered by complementary notions' (*NES*, 146). It is matter of progressing through new perspectives, rather than being pushed forward by a single dominant one.

Dialectics then becomes the questioning search for 'variations under identity' that 'shed light on the first thought by completing' (*NES*, 150). This is not, then, a structure of scientific revolutions. It is a structure of scientific additions and revisions, where each addition involves changes in the objective status of what it adds to. This is shown in a very beautiful passage on laws: 'We shall not speak of simple laws that are then disrupted, but of complex and organic laws touched sometimes by certain viscosities, certain obliterations. The old simple law becomes a simple example, a mutilated truth, the beginnings of an image, a sketch on a board' (161).

However, if Bachelard's account of discontinuity raises problems for Deleuze, this is even more the case for DeLanda's interpretation of Deleuze's ontology. This is because this account of ontology has neither a lasting and thorough role for doubt, nor for rectification. These have a place once, at the outset, when alternative ontology is rejected, but thereafter a final model takes sway and cannot be questioned in a radical manner: 'This will complete the elimination of essences we have begun here, ensuring that multiplicities possess their own historicity and preventing them from being confused with eternal archetypes' (*ISVP*, 80–1). It is the notion of a complete elimination of anything that is troubling from Bachelard's point of view (and I hope to show also from Deleuze's), since Bachelard's philosophy of science follows close on the heels of the shock of the overthrow of an apparently stable theoretical and explanatory system. The difference can be summed up in terms of the contrast between the shock of the possibility of error in a theoretical system that had proven to be highly successful and the delight at arriving at an even more powerful and, from DeLanda's point of view, a more philosophically attractive one. A further sign of this difference can be found in the styles of reflection and enquiry in all three positions where Deleuze and Bachelard give much greater space to questions and to critical distinctions, whereas DeLanda opts for explanation and a search for theoretical completeness.

The place and significance of scientific progress

Deleuze denies the possibility of progress as defined by Bachelard in *Le nouvel esprit scientifique*. As Deleuze argues in *Difference and Repetition*, there can be actual progress, but this must be set against the eternal return of difference, that is, the significance of actual progress changes in terms of the eternal and necessary return of intensities, virtual ideas and sensations. So it is perfectly possible to speak of scientific progress, but it is not possible to assign any independence for that progress with respect to value and to sensibility. In other words, the status of scientific progress cannot be separated from other values that may well contradict claims to improvement. Philosophy cannot be simply progressive in the same way as science.

The second key point to be made by Deleuze against Bachelard concerns continuity. Where Bachelard claims that dialectics is a matter of the search for completeness through discontinuous but not opposed terms, Deleuze argues that beneath every actual difference, beneath every disparity, we find a continuous transcendental condition for actual difference, where actual difference is defined in terms of identity. In *Difference and Repetition*, this argument is developed in terms of pure differences underlying measured spaces (*DR*, 229). These points are linked, since when Deleuze speaks of the eternal return of difference he means the return of the expression of continuous virtual ideas and intensities; under different configurations or perplications (envelopments), in the case of intensities, and different relations of distinctness and obscurity, in the case of transcendental ideas. There is no possibility of discontinuity between intensities and ideas — such breaks only appear when they are actually expressed. Even then, a complete expression must always take account of the connection of all actual things through the virtual (252).

This connection negates the possibility of a view of time as a linear or cyclical progression, since the condition for this progression is a continuous realm that cannot be satisfactorily mapped on to a linear account. Underlying actual things and standing as transcendental conditions for their variation, we find intensities and Ideas that cannot be finally separated from one another, because such separation would introduce illegitimate limits in the conditioned realm of actual things. So Ideas and intensities can only be determined as separate according to degrees and according to relations of distinctness *and* obscurity, where distinctness only appears with a wider varying set of more or less obscure relations.

Bachelard's counter to such claims is developed in *La dialectique de la durée*.[7] It is constructed around the thesis that psychology and phenomenology of time imply discontinuity. This discontinuity of time implies an ontological discontinuity. Every continuity is therefore illusory and, in fact, secondary with respect to a dialectics that comes out of the possibility of affirmation or negation in activity: 'To think is to abstract certain experiences, it is to plunge them into the shadow of nothingness. If one objected to us that that these effaced positive experiences subsist nonetheless, we would answer that they subsist without playing a role in our actual knowledge' (*DD*, 16). Because we make negative judgements we have to suppose that reality is discontinuous, in the sense of allowing gaps or empty space and time in existence.

For Deleuze, this is to miss the role of passive syntheses as transcendental conditions for activity. These conditions imply continuity at the level of Ideas and a complicated continuity at the level of time. But a further rejoinder can be found in Bachelard's response to the question of the transcendental: 'Time is therefore continuous as possibility, as nothingness. It is discontinuous as being' (*DD*, 25). Time as condition must be discontinuous, given the nature of our consciousness of things as having the potential to exist or not. There is a discontinuity between these two states, and Bachelard will go on to argue that this discontinuity is replicated in consciousness through our power to negate and to make decisions between different possible routes.

This difference between Deleuze and Bachelard on time and continuity allows for a further set of questions to be put to DeLanda. This time, they are not about science in terms of falsifiability, but in terms of historicity. Against Bachelard's position, DeLanda refuses a linear account of progression for the same reasons as Deleuze: the different times implied by the relation of the virtual to the actual. But DeLanda still retains historicity as the key concept for understanding Deleuze's work on time: 'In other words, in Deleuzian ontology there exist two histories, one actual and one virtual, having complex interactions with one another' (*ISVP*, 155). This is a return to notions of the possibility of progress — albeit very complex — dependent on the relations between the two realms and within each. But this would not be the case were one realm not historical at all, a point that Deleuze's critique of Bachelard depends upon. So a counter to DeLanda would be that we do not have a system of two histories, but a relation between histories and something that always resists mapping out in terms of history, because it is the condition for different actual histories and ways of understanding or explaining time, not as a limited or fixed set, but as a set that is radically open and prone to new accounts. If there were simply two histories that produced complex and changing relations, these could still be formalized in such a way that minimal accounts of emergence and progress would become possible (we have seen this in DeLanda's reliance on narratives of error and trends towards truths that avoid such errors). But this misses the point that Deleuze requires a metaphysics where no such formalization and restriction can be given. His philosophy is not about *a* time (or times), however complex, but about the conditions for many actual times and many more *new* actual times.

Syntheses of time and Deleuze's critique of discontinuity

The opposition around continuity described above can be summed up through two opposed arguments. Deleuze's line is that actual events presuppose a transcendental continuity, because such events cannot simply be accounted for in terms of identities. Identities are encountered in events that vary according to a 'drama' of multiple sensations and hence intensities. For example, there is no finally isolated moment of decision in psychology, only the awareness of deciding or of signs of a decision, abstracted from endless and variegated rises and falls in tension in feelings and processes. The abstraction can make us think that time is essentially discontinuous, in the sense of thinking 'Everything changed here, at this point'. But the break is always changing in significance according to the variations that surround it, as in a declaration and its context, for example. For Bachelard, on the other hand, the argument goes that science and phenomenology are the only proper sources of evidence for conclusions concerning time. What they show is that we have to assume that time is discontinuous, in order to account for the psychology of decisions and choices, and for the phenomenology of intentionality. Were time continuous, then our sense of points of decision, of breaks where events could go one way or another, would be mistaken. Our experiences are of discontinuous events, where things stop and start, where they can be made to stop and start, and where our directedness to events presupposes such breaks. The fact of negation, where we can stop things, or where things stop, or where our directedness implies the necessity of stops and starts, shows that time must be discontinuous.

In reply, we can look at a scientifically isolated point of decision — or at least, what stand as *signs* of decision — such as 'The rat moves to A' or 'The mapping of brain patterns changes dramatically at this point'. From a Deleuzian point of view, each isolation is open to re-examination according to wider patterns of significance, that is, according to different flows of sensations and of the problems that surround them. In other words, the scientific or phenomenological point will move. It may even be located at plural points, or according to neighbourhoods or stretches that deny the importance of a single point.

But this shift to relative plurality is not the issue for Deleuze. Rather, it is what the shift presupposes that bothers him — not what it leads to. Why did we change our view of what stands as the point of decision? Can that change be explained in purely scientific terms (new discoveries) or phenomenological terms (different and more complex views of intentionality)? Or, rather, do we have to look further in terms of why we continue to search for changes in a given direction? Shouldn't we look at why that direction changes and at why we view the results of that change in different and perhaps ultimately individual ways?

Deleuze's transcendental deductions around three syntheses of time point to conditions for discontinuity that are themselves continuous. An active decision presupposes all the repetitions and variations that have come together into forming a being capable of making the decision. Indeed, even the short-hand of 'being' or 'actor' is insufficient. What we should say is that a given situation, comprising a decision and its environment, has no limit in principle with respect to the extent of the environment and actor in terms of antecedents (*DR*, 77). It is possible to say that an ancestry 'decides' or that the decision lies in the relation between a climate and that ancestry. It is not possible to say that any point of that ancestry or environment is excluded in principle. Furthermore, actual repetitions that lead to a decision have to be extended into a field of virtual memory (Bergson's pure past). It is not only the hard-wired aspects of the past that matter, but also soft and highly variable virtual ones (81-2). This is something that Deleuze shows in his work on cinema. Our memory is a swirling and changeable record of the past. Yet it is played out in present acts. Again, no limitation in principle makes sense. More seriously, there cannot even be a linear limitation, as might have been thought in the actual repetitions ('This must have happened first'). Memory can be re-jigged in the present, to the point where we have to say that we act on the virtual past and also, therefore, on the relation of that virtual present and its relation to actual past repetitions. Equally, though, the virtual past acts on us, thereby setting off relations of reciprocal determination of memory and actuality.[8]

But does any of this work on memory imply continuity? Should we still not speak of actual identifiable things in repetitions and environments? Should we not do the same with respect to memories? Everything may have to be thought of in limitless chains. This may force us to accept the necessity of contingent abstractions. But that does not mean that such abstractions do not take place in essentially discontinuous realms.

To answer these questions we have to go beyond Deleuze's first two syntheses of time ('Every present event presupposes syntheses of actual chains of repetitions' and 'Every present act presupposes the synthesis of the whole of the virtual past'). However, prior to that, it is important to make a point concerning the relations that hold between all three of Deleuze's syntheses. Each one is incomplete without the others — they presuppose one another. We have already seen that the first synthesis and the second are connected and cannot be separated. This is because actual repetitions and acts take place and acquire significance within virtual memory. This is as true for a material or biological process as it is for a conscious being. Deleuze's point is that the synthesis of the pure virtual past is a necessary condition for the synthesis of an active present: 'There is thus a substantial temporal element (the Past which was never present) playing the role of ground' (*DR*, 82). That ground is what allows an act in the present to be determined in terms of how it rearranges the whole of the past and hence also in terms of how it acquires significance, determinacy and value in relation to the present: 'if the new present is always endowed with a supplementary dimension, this is because it is reflected *in* the element of the pure past in general, whereas it is only *through* this element that we focus upon the former present as a particular' (82). The present is incomplete unless is considered in relation to all the things it passes away into (the pure past).

It is possible to think of this interdependence in terms of value. Seen as brute material processes, chains of repetitions are neutral with respect to value (Why celebrate the birth of an animal? Shrug at the erosion of a rock? Ignore a microscopic change?). When events are selected, value impinges to introduce hierarchies. But what is this value? For Deleuze, it is itself a selection through sensations and these depend on past associations of ideas and sensations. There is therefore a virtual, immaterial, trace of selections that runs through all of the virtual past and this trace introduces value and selection into actual processes. There is a virtual history of value that allows for determinations in the actual. Why did you care so about that rock face? What selections did that care imply?

This reference to selection as the appearance of the new in an unfolding series of events is at the core of both Deleuze's and Bachelard's arguments. For Deleuze, selection implies a third synthesis of time as a relation to the future. This synthesis implies continuity in all syntheses and through all things. A continuous relation — that Deleuze will define in terms of intensities and virtual Ideas — is presupposed by all actual events. For Bachelard, it is quite the opposite: selection and the new presuppose discontinuity.

Deleuze's argument is that selection, as the drive towards the new, presupposes a cut, assembly and transformation of all of time. The first point is not controversial, at least in this context, since it supports Bachelard's point. To select the new, we presuppose that it cuts away from the past in some way. The next point is that though there is a cut, it is one that takes place within the backdrop of the past. Therefore, the cut projects that past into the future. A decision or an unconscious selection does not only cut away from the past, it brings something new into the past and brings the past into the future. So though there is a break, there is also an assembly. A discontinuous and continuous time are implied by selection. But isn't this a contradiction? How can time be both continuous and discontinuous? Is it not rather the case that time is discontinuous throughout and that Deleuze's assembly is an assembly of prior cuts and later ones? His answer lies in the third property of selection. The assembly of the past and the future are transformations of them. So it is possible to speak of a cut and of an assembly, because the assembly is of different things. When thinking of the future (F) as different from the past (P), we may be tempted to think that the difference lies between P and F. But Deleuze's point is that in a selection we move from an assembly P/F to a new assembly P′/F′. We select a new past and a new future. So any difference is between P/F and P′/F′. However, is this not even more nonsensical than the previous contradiction? How can we change the past and the future in the present?

In *Difference and Repetition*, Deleuze's answer is usually couched in terms of Nietzsche's doctrine of eternal return, but I want to give a different version that links more easily to his ideas about intensities and virtual Ideas. It is the case that any given actual identical thing cannot return, what returns are pure differences and what changes is the relation of these to actual things. When we spoke of the pure past and of virtual Ideas and intensities earlier, these could have been understood as identifiable memories — open to representation. For Deleuze, the virtual is the transcendental condition for transformations, that is, for the sensation that something is actually different though in an unidentifiable way (if it could be identified, then it would not be new in the sense of implying a cut).

These conditions are always defined as continuous for Deleuze. Otherwise, forms of identity and representation would return in the virtual, thereby contradicting his argument that the new must be radical in the sense of departing from the present and from the past, whilst transforming them: 'The synthesis of time here constitutes a future which affirms at once both the unconditioned character of the product in relation to the conditions of its production, and the independence of the work in relation to its author or actor' (*DR*, 94). For this unconditioned character to hold, and yet for there to be an assembly and transformation of the conditions of production, the new presupposes something that escapes both the actor and the production (past and present). This is the transcendental field of the virtual. The radical nature of the new as expressed through sensations implies therefore that this field must be continuous — and hence independent of actor and production, in the sense of in principle unidentifiable in terms of them. It must change only as continuous, that is, in terms of relations of distinctness and obscurity, rather than in terms of relations of opposition and identity.

The transcendental field is a continuous multiplicity of varying relations that stands as the condition for the new as cut and transformation, for example, as condition for the fractured I and dissolved self: 'As we have seen, what swarms around the edges of the fracture are Ideas in the form of problems — in other words, in the form of multiplicities made up of differential relations and variations of relations, distinctive points and transformations of points' (*DR*, 259). For Deleuze, the new presupposes continuity; otherwise, we could not explain its novelty.

Conclusion: the cost of abandoning the transcendental

Deleuze and Bachelard develop arguments for different philosophical dialectics in response to the requirement to take account of radical novelty. Scientific revolutions are paradigms for this kind of novelty and each thinker is careful to situate any given science within a philosophical frame that allows for abandonment or radical refinement of an established position in favour of the addition of new ones or a straightforward supplanting of old by new. The two thinkers disagree, though, on the relative priority of accounts of continuity and discontinuity underlying philosophical definitions of reality and of thought in relation to reality. For Deleuze, actual worlds are discontinuous, but presuppose a virtual continuity. For Bachelard, reality is discontinuous, but we have to construct dialectical continuities that relate discontinuous elements. For example, scientific revolutions are accounted for in Deleuze's philosophy because any science must present a necessarily limited and in principle replaceable theoretical representation of actual things; yet the condition for this limitation is a transcendental continuity of a very special kind, that is, a carrying multiplicity of Ideas that resist identification but that must be expressed in many different actual theories and representations. In Bachelard's work, the assumption that no theory can be finally settled and is open to change is a necessary reflection of scientific practice and of the phenomenology of thought; doubt and questioning are necessary for scientific practice and essential for thought.

The problem for Delanda's science-based account of Deleuze's philosophy, despite its great explanatory power, is that it underestimates the need for a principled philosophical openness in light of scientific history and practice. This leads to a devaluing of the transcendental move in Deleuze's work and to a science-based definition of terms such as multiplicity and the virtual, when they cannot be simply determined in this way without inviting grave problems with respect to historical situation. How can we know that any given science is definitive or even on the way to being definitive? The next discovery on, say, dark matter may well challenge our current cosmology to the point where it can be seen as obsolete or wrong in fundamental ways. It is crucial for philosophy, if it is to retain some independence from science, that it be capable of explaining and responding to both the possibility of change and the practice of ongoing efforts to falsify and confirm scientific theories. If philosophical terms are seen as merely derived from a given science, as for example in DeLanda's definitions of Deleuze's extensive spatium as derived from thermodynamics (*ISVP*, 158–9), then this independence and awareness will have been compromised. It is not that DeLanda's interpretation is 'wrong' — though it certainly conflates virtual and actual in ways that conflict with Deleuze's transcendental deductions. It is rather that it risks underplaying the way in which Deleuze inherits a strong scepticism, from Hume, and a powerful and primary sense of transcendental condition, from Kant. A possible answer to this point could be that it is necessary to run this risk in order to release the constructive power of Deleuze's thought, that is, in order to connect it to contemporary science in its current applications. Yet this invites the riposte that for Deleuze *any* science must betray its condition within a virtual continuity when it offers a particular explanation and becomes allied with a particular technology. The distinction DeLanda makes between Royal and minor science (*ISVP*, 179–80), following Deleuze and Guattari in *A Thousand Plateaus*,[9] is irrelevant here, since both models of science offer explanations that restrict future ones through relative identifications, irrespective of whether one model is totalizing and representational and the other relative and practical.

NOTES

1 There is a much longer discussion of the differences between Deleuze and Bachelard that forms the background to the work here in my *The Transversal Thought of Gilles Deleuze: Encounters and Influences* (Manchester, Clinamen, 2005). Extended material on Deleuze and Bachelard, but with no application to DeLanda can also be found in my 'How radical is the new? Deleuze and Bachelard on the Problems of Completeness and Continuity in Dialectics', *Pli, the Warwick Journal of Philosophy* 16 (2005), 149–170.

2 See, for example, Mark Bonta and John Protevi's discussion of complexity theory and of DeLanda's work in *Deleuze and Geophilosophy: a Guide and Glossary* (Edinburgh, University Press, 2004), 16–21.

3 Manuel DeLanda, *Intensive Science and Virtual Philosophy* (London, Continuum, 2002), 4. Henceforth *ISVP*.

4 Gilles Deleuze, *Empiricism and Subjectivity*, translated by Constantin V. Boundas (New York, Columbia, 1991 [1953]).

5 Gaston Bachelard, *Le nouvel esprit scientifique* (Paris, Presses Universitaires de France, 1934). Henceforth *NES* (my translations); *The New Scientific Spirit*, translated by A. Goldhammer (Boston, Beacon Press, 1985).

6 See Deleuze's study of the works of Curie in *Difference and Repetition*, translated by Paul Patton (London, Athlone, 1994), 286 (henceforth *DR*) and my discussion of Deleuze's account on explanation in *Gilles Deleuze's Difference and Repetition: a Critical Introduction and Guide* (Edinburgh, Edinburgh University Press, 2003), 168–71.

7 Gaston Bachelard, *La dialectique de la durée* (Paris, Presses Universitaires de France, 1950). Henceforth *DD* (my translations); *Dialectic of Duration*, translated by Mary McAllester (Manchester, Clinamen, 2000).

8 See Deleuze's study of these processes in the films of Orson Welles in Gilles Deleuze, *Cinema 2: The Time-Image*, translated by Hugh Tomlinson and Robert Galeta (Minneapolis, University of Minnesota Press, 1989), 105–16, especially 110–11.

9 *A Thousand Plateaus: Capitalism and Schizophrenia*, translated by Brian Massumi (London/Minneapolis, University of Minnesota Press, 1987), 367–74.

The Difference Between Science and Philosophy: the Spinoza-Boyle Controversy Revisited

SIMON DUFFY

Abstract

This article examines the seventeenth-century debate between the Dutch philosopher Benedict de Spinoza and the British scientist Robert Boyle, with a view to explicating what the twentieth-century French philosopher Gilles Deleuze considers to be the difference between science and philosophy. The two main themes that are usually drawn from the correspondence of Boyle and Spinoza, and used to polarize the exchange, are the different views on scientific methodology and on the nature of matter that are attributed to each correspondent. Commentators have tended to focus on one or the other of these themes in order to champion either Boyle or Spinoza in their assessment of the exchange. This paper draws upon the resources made available by Gilles Deleuze and Felix Guattari in their major work *What is Philosophy?*, in order to offer a more balanced account of the exchange, which in its turn contributes to our understanding of Deleuze and Guattari's conception of the difference between science and philosophy.

Keywords: Spinoza, Boyle, Deleuze, scientific methodology, rationalism, experimentalism, plane of immanence, plane of reference

A number of studies have been devoted to the examination of the correspondence that took place between Benedict de Spinoza (1632–77) and Robert Boyle (1627–91), by means of the intermediary Henry Oldenburg, during the period from 1661 to 1663. It was upon the instigation of Oldenburg that Spinoza was made aware of Boyle's then recently published book *Certain Physiological Essays* (1661).[1] Boyle was a founding member of the Royal Society (established in 1644), and is considered to be one of the leading English natural philosophers of the Scientific Revolution. He was the first to style his type of natural philosophy as the corpuscular philosophy, which brought together the role of particulate matter in the explanation of natural phenomena with the dual mechanical principles of matter and motion. Oldenburg, the then secretary to the Royal

Society, had met Spinoza during a trip to Holland in the summer of 1661. The correspondence contains Spinoza's responses to a number of experiments on nitre, and on solidity and fluidity, that Boyle gives details of in the *Essays*.

The two main themes that are usually drawn from the correspondence and used to polarize the exchange are the different views on scientific methodology and the nature of matter that are attributed to each correspondent. Commentators have tended to focus on one or the other of these themes in order to champion either Boyle or Spinoza in their assessment of the exchange. This paper draws upon the resources made available by Gilles Deleuze and Felix Guattari in their major work *What is Philosophy?*,[2] in particular their distinction between science and philosophy, to offer a more balanced account of the exchange. Those commentators who champion Boyle invariably offer a scientific assessment of the exchange that relies upon a general presentation of the limitations of rationalist philosophy, and by extension Spinoza's philosophy, to provide the grounds for the development of an adequate scientific methodology.[3] Whereas those who want to balance the ledger by offering a philosophical defence of Spinoza make a lot of the inconsistencies in Boyle's natural philosophy in comparison to Spinoza's metaphysics.[4]

Neither of these approaches is up to the task of providing an adequate assessment of the exchange. On the one hand, as Peter Anstey notes, 'Boyle was adamant' that his mechanical philosophy 'was a theory and not a set of metaphysical first principles upon which a science of nature was to be based'.[5] To find Boyle inconsistent in this respect is to misconstrue the nature of his project. Indeed, Boyle had a 'self-confessed aversion to system building' and was not interested in systematizing the corpuscular philosophy.[6] On the other hand, those who champion Boyle contrast the importance of Boyle's experiments to the development of a properly empirical scientific knowledge with the metaphysical principles of Spinoza's mechanical physics. From this point of view, it is tempting to regard the correspondence between Spinoza and Boyle as providing a particular case of the opposition between a quintessential rationalism and an emergent experimentalism. The concept of rationalism in this instance being characterized as the exclusive doctrine that knowledge is deduced from principles that are determined independently of experience, or at least such that they have a priority over it. And experimentalism being characterized conversely as an empirical doctrine that advocates the use of experimental methods in determining the validity of ideas, the principles of which are hypotheses. According to such a reading, Spinoza is criticized as being the pure philosopher who refuses to admit hypotheses that put rational principles into self-contradiction, and Boyle is championed as the pre-positivist scientist who refuses to reason beyond the facts and relations that are determined by experimentation. This sets up a radical antagonism between two mutually exclusive positions, both in terms of the particular theses that they each develop, and in terms of the theoretical conception that they each have of their own work.[7] Elkhanan Yakira can be thanked for judiciously pointing out that such an interpretation is 'both simplistic and excessively limited'.[8] The distinction between them is only problematically schematized as that between a quintessential rationalist and an experimentalist championing the new science. If the exchange that took place between Boyle and Spinoza is examined with the due attention that it deserves, it should become clear that the points of disagreement in the correspondence are of much less importance than a superficial reading of the correspondence would at first suggest.[9]

The characterization of Spinoza as a typical rationalist philosopher risks obscuring Spinoza's own engagement with, and development of, the new mechanistic science, particularly when it comes to the very empirical and constitutive nature of Spinoza's first kind of knowledge, or the imagination, and its importance for the development of reason, or the second kind of knowledge. Just as it is problematic to presume that Boyle's speculations about the textures of particles,[10] which are of singular importance to the kinds of interpretations that he gives of his experiments, are vindicated by subsequent developments in chemistry. Indeed, there are certain moments in the correspondence that could have given Spinoza concern for the adequacy with which Boyle was able to distinguish the trajectory of his scientific endeavours from the Scholastic tradition that both of them were keen to move beyond. Rather than attempting to assess the distinction between the two correspondents as being characteristic of the distinction between two different kinds of philosophy — that is, between Spinoza's particular brand of rationalism and the kind of empirical natural philosophy that Boyle's experimentalism implies — or between the two different scientific methodologies that can be extracted respectively from these positions, I propose to assess the distinction from the point of view of the question, already implicitly problematized, of the relation between science and philosophy.

Boyle and Spinoza were corresponding at the dawn of a new scientific era, and Boyle can be characterized 'as a transitional figure in the parting of ways for philosophy and science'.[11] The limited and episodic nature of the exchange provides the focus for an examination of this relation precisely at the time when science and philosophy began to distinguish themselves as separate disciplines. The account of the difference between science and philosophy that is given by Deleuze and Guattari in *What is Philosophy?* offers new resources to examine the controversial exchange between Spinoza and Boyle, and this exchange in turn provides the resources for an explication of Deleuze and Guattari's understanding of the difference between science and philosophy.

In *What is Philosophy?*, Deleuze and Guattari define philosophy as 'the creation of concepts' (*WIP*, 41) which enter into resonance with one another on what they define as 'a plane of immanence' (35). Philosophy is therefore 'at once concept creation and instituting of the plane' (41), that is, philosophy both creates concepts and generates the plane on which they are then distributed. Unlike philosophy, the object of science, for Deleuze and Guattari, 'is not concepts but rather functions that are presented as propositions in discursive systems' (117). Like concepts, functions also need to be created, and once created, they are distributed not on planes of immanence, but rather on planes of reference.

For Deleuze and Guattari, the concept has nothing to do with representations nor with propositions: it is not simply the idea of a form, or a container for cognitive content, nor is it simply a bearer of truth-values. Instead, they argue, 'All concepts are connected to problems without which they would have no meaning and which themselves can only be isolated or understood as their solution emerges' (16). A concept therefore only has meaning as a function of the problem 'that it resolves or helps to resolve' (79). The meaning that the problem confers on the concept is not the immediate signification of a proposition. The problem should not be reduced to the case of a solution which immediately exhausts the instance of the problem, nor should it be confused with the ordinary process of putting 'the same subject of a thesis into contradiction'.[12] Rather than determining the problem by the assurance of a solution, the problem should be understood to be determined by the urgency of its related questions. Therefore, rather than moving from problems to solutions, Deleuze and Guattari propose moving from the problem to what conditions the problem 'and forms the cases that resolve it'.[13] Problems can therefore be understood to generate new modes of questioning, opening different perspectives on the more familiar or conferring interest on the given that until then had remained insignificant.[14] The components of a concept are those other problematic or fragmentary concepts that have been created in the past that are reoriented in relation to one another by virtue of a particular problem. The concept as such is a point of condensation, or neighbourhood, which governs and gives consistency to these heterogenous components.[15] Deleuze and Guattari describe concepts as 'centers of vibrations, each in itself and every one in relation to all the others. This is why they all resonate rather than cohere or correspond with each other' (*WIP*, 23). The concept is a point or 'state of absolute survey (*survol*) in relation to its components', which are 'traversed by it at infinite speed' (21). In this way, the concept is 'immediately co-present to all its components' (20). The infinite speed with which the concept traverses all of its components in a state of absolute survey is retained from the virtual, which is characterized by Deleuze and Guattari in *What is Philosophy?* as 'chaos'. This is not the chaos of chaos theory, as Deleuze and Guattari argue:

> Chaos is defined not so much by its disorder as by the infinite speed with which every form taking shape in it vanishes. It is a void that is not a nothingness but a virtual, containing all possible particles and drawing out all possible forms, which spring up only to disappear immediately, without consistency or reference, without consequence. (118)

The example that Deleuze and Guattari draw upon for their characterization of chaos is the crystallization of a superfused liquid.[16] It is in the very relation of resonance, co-presence, or absolute survey that the infinite speed with which forms take shape and vanish in chaos is retained in philosophy, and that the virtual (chaos) is given consistency. It is this very consistency that constitutes the plane of immanence of a particular concept, and that determines a particular 'image of thought' in relation to that concept. As Deleuze and Guattari note: 'The Plane of immanence is not a concept that is or can be thought but rather the image of thought, the image thought gives itself of what it means to think, to make use of thought, to find one's bearings in thought' (37).

Deleuze and Guattari argue that 'every great philosopher lay[s] out a new plane of immanence' (51), such that 'there are varied and distinct planes of immanence that (...) succeed and contest each other in history' (39). Indeed they argue that the history of philosophy 'exhibits so many quite distinct planes' that 'we can and must presuppose a multiplicity of planes' (50). The actual practice of philosophy, according to Deleuze and Guattari, consists above all in disengaging problematic or fragmentary concepts that have been created by philosophers of the past — along with the plane of immanence that each concept generated and on which it is distributed — from the supposed order of succession of achieved philosophical systems, by grafting these concepts (of the past) onto one another to create new concepts. This is done by selecting those concepts (of the past) which enter into reciprocal relation or are problematized with the signature concepts of other philosophers,[17] thereby extending each of the associated planes of immanence by determining their actual coexistence in relation to one another.

Deleuze and Guattari do not consider the history of philosophy to be a succession of achieved philosophical systems. Instead they characterize a time of philosophy that is rather 'a grandiose time of coexistence that does not exclude the before and after but *superimposes* them in a stratigraphic order' (59). The 'time of philosophy' is therefore not the subject of a future (after) and a past (before) — Scotus before Descartes, Spinoza after Descartes — but rather constitutes a 'stratigraphic time where "before" and "after" indicate only an order of superpositions' (58). Duns Scotus can therefore come 'after' Descartes, insofar as the Scotist plane of immanence, on which the concept of the univocity of being is distributed, is superimposed on the Cartesian plane of immanence, on which the concept of substance dualism is distributed.[18] These concepts are selected and grafted together by Spinoza to create his concept of substance monism. Deleuze and Guattari maintain that 'very old strata can rise to the surface again, can cut a path through the formations that covered them and surface directly on the current stratum to which they impart new curvature' (58). Philosophical concepts therefore do not enter into the linear progression of an evolutionary history, but are rather distributed over a plane of immanence that generates instead an intensive temporality of its own. It is the process of the selection of problematic or fragmentary concepts (of the past) that determines the creation of new concepts and it is the operation of this process that Deleuze and Guattari consider to be characteristic of the practice of philosophy.

Problems subtend not only the creation of concepts, but also the creation of functions. However, while concepts conserve the infinite speed of the virtual (chaos), a function effects a 'fantastic slowing down' (118). 'To slow down is to set a limit in chaos', to limit the speed with which forms take shape and vanish in chaos. Deleuze and Guattari argue that 'it is by slowing down that matter, as well as the scientific thought able to penetrate it with propositions, is actualized' (118). What science does in response to problems is to substitute a horizon for the infinite, in order to determine solutions. It does this by giving reference points to the virtual (chaos), reference points such as the universal constants of physics: the invariance of the speed of light, Plank's constant, absolute zero, etc. And it is with these reference points, or 'limits', that science constructs functions. '[Science] relinquishes the infinite, infinite speed, in order to gain *a reference able to actualize the virtual*' (118). Science actualizes the virtual (chaos) through functions. A function is composed of elements, what Deleuze and Guattari call 'functives', which include these limits and the variables that they are determined in relation to. A function therefore consists of a functive regarded in its relation to one or more other functives in terms of which it may be expressed. This relationality determines a state of affairs that actualizes the virtual on a plane of reference and in a system of co-ordinates. As they note: 'reference is a relationship (. . .) of the variable with the limit' (118–19). The plane of reference of science is a discursive system in which functions are presented as propositions determinative of states of affairs. The plane of reference is not unitary, but is instead constituted by the irreducible, heterogenous systems of co-ordinates that each limit generates on its own account (119–20). The meaning that a problem confers on a function is the state of affairs in which it insists.

Deleuze and Guattari also reconceptualize the history of science, characterizing instead a time of science that 'is not confined to a linear temporal succession any more than philosophy is' (124). But, instead of a stratigraphic time, which expresses before and after in an order of superimpositions, they maintain that 'science displays a peculiarly serial, ramified time, in which the before (the previous) always designates bifurcations and ruptures to come, and the after designates retroactive reconnections' (124). One of the results of this is a different way of conceptualizing scientific progress. Scientists' proper names, when understood according to this other time, mark 'points of rupture and points of reconnection' (125) in the ramified time of scientific development.

The correspondence between Spinoza and Boyle can be construed as being played out across these two quite different planes: the plane of immanence of philosophy; and the plane of reference of science. On the one hand, there is an interrogation of the emerging domain of corpuscular chemistry that turns around the problem of the nature and composition of nitre, an interrogation that is properly speaking philosophical for Spinoza, but that remains scientific for Boyle. On the other hand, there is a specific interrogation between the two protagonists, at the level of scientific methodology, of the adequacy of the particular experiments being carried out to this end. The difference between the different positions taken up by each of the protagonists in their correspondence, as we shall see, gives an exemplary expression of the difference between the plane of reference of Boyle's corpuscular chemistry and the plane of immanence of Spinoza's metaphysics.

Mechanics was the emerging scientific discipline of the time that became the general principle of scientific explication and practice, what Deleuze and Guattari would describe as one of the planes of reference for the emerging scientific enterprise. Spinoza, as much as Boyle, adopts the mechanical principles as a method of explication. While Spinoza incorporates them into his philosophy alongside the principles of his metaphysics, the mechanical principles that Boyle uses to support a corpuscular world view are determined by hypotheses, and the ultimate criteria of the veracity of hypotheses is experimental evidence. Even though such a mechanics promised to provide a comprehensive heuristic structure for the explanation of all corporeal phenomena, Boyle was committed to a number of speculations about corpuscular entities that at the time remained beyond its explanatory resources.[19] These included the strong 'empirical sense that some substances were in a chemically significant way permanently distinct from other substances', for example the end products of chemical analysis, for which Boyle's experiments on nitre will function as an example. The theoretically equivalent speculation being 'that some mechanical textures of particles possessed a characteristic quality of being indisruptable (*sic*)'.[20] It was only much later, with the development of analytic chemistry towards the end of the eighteenth century, that such speculations were adequately reformulated with the postulation of entities and qualities that were unfamiliar to Boyle, and whose existence was then able to be demonstrated experimentally. Macherey quite correctly warns against being tempted to loan to Boyle the conceptions of analytic chemistry that are radically different to those of his corpuscular chemistry.[21] Indeed, Boyle's 'theory of matter is now completely debunked'.[22]

The distinction between different particles of matter that Boyle introduces looked forward to the notion of a chemical element, which only really appeared much later with Lavoisier (1743–94). In the context of Boyle's corpuscular chemistry, such an element would be understood simply as an elementary body, or particle, composed of the particulate matter that was the ultimate material constituent of all corporeal objects.[23] This distinction could be understood to identify in nascent form the notion of a particle of matter having irreducible chemical properties which are not lost when it is integrated with another to create a compound matter. However, Boyle's speculations about the non-disruptable quality of some textures of particles and Lavoisier's notion of the chemical element are quite different. Lavoisier clarified the concept of a chemical element as a simple substance, rather than a corpuscular object, that could not be broken down by any known method of chemical analysis, and he devised a theory of the formation of chemical compounds from elements. He also compiled a list of elements, which included oxygen, nitrogen, hydrogen, phosphorus, mercury, zinc, and sulphur, that formed the basis for the modern periodic table of chemical elements, developed by Mendeleev in 1869. So not only did they not have a conception of a chemical element, but also neither Spinoza nor Boyle could have had a conception of chemical reaction in the sense that it takes after Lavoisier.

The speculations of Boyle's corpuscular chemistry designate the bifurcations and ruptures to come in the emerging discipline of chemistry that subsequently led to the development of the plane of reference that is characteristic of analytic chemistry. It is important to note however that it is only retroactively that this plane of reference is reconnected with Boyle's speculations. To read the relation between the two as having been linearly, or serially, determined, that is, as if Boyle's speculations lead directly to the development of the notion of chemical element, would be to overdetermine Boyle's contribution to the development of analytic chemistry, and to succumb to the temptation that Macherey warns against. According to the ramified time of science, the connection between Boyle's speculations and the developments in analytic chemistry are rather made retrospectively, from the latter to the former.

According to the principles of mechanics that both Boyle and Spinoza adopt and develop, the composition of different bodies, in terms of shape and movement, or material textures, is determined mechanically by the arrangement of their constituent particles, themselves without quality, however these arrangements are manifested to the senses in the form of qualities. Boyle's speculations do seem to imply an intrinsic nature proper to particles that differs from one particle to another, the implication being that matter has qualities that are not solely reducible to the movement of particles. This would contradict Spinoza's mechanism of the second part of the *Ethics* where he states that 'Bodies are distinguished from one another by reason of motion and rest, speed and slowness, and not by reason of substance'.[24] However, Boyle doesn't actually give a clear indication of his view of the nature of matter until after the correspondence,[25] and even then his position remains inconsistent. So, while not furnishing us with a definitive view as to the ontological status of the sensible qualities, Boyle recontextualizes the discussion of many of the central issues that pertain to the nature of the sensible qualities within the development of his corpuscular chemistry.[26] The question, however remains: Do Boyle's speculations make a strong enough claim to the stability of an intrinsic nature proper to different particles to cause Spinoza concern?

In his first letter to Boyle, Spinoza argues that it is not necessary to examine whether the demonstrations that Boyle offers to show that 'the tangible [or sensible] qualities depend only on motion, shape, and the remaining mechanical affections' are 'completely convincing' because 'he does not present these demonstrations as Mathematical' (Letter VI). Whatever speculative content there is to Boyle's understanding of the nature of the sensible qualities, this comment indicates that Spinoza is not overly concerned by it, because he does not consider Boyle to have demonstrated this with any rigour, which is what Spinoza's reference to mathematics serves to indicate. So contrary to the above suggestion that this speculation on Boyle's part would contradict Spinoza's mechanism, Boyle's speculation about the stability of these qualities, or the non-disruptability of certain characteristic qualities at a chemical level, does not necessarily imply a difference in intrinsic nature between the particles exhibiting such qualitative stability. In fact, Spinoza's distinction between fluid, soft and hard bodies in the Lemmas of the second part of the *Ethics* would have provided support for such speculation.[27]

It is the very transitional nature of the corpuscular chemistry which Boyle was in the process of developing that leaves his comments in the correspondence open to criticism of this sort. As it turns out, the speculative component of Boyle's train of thought proved justifiable, despite being based on a theory of matter that was proved incorrect. This could, however, only be determined retrospectively, from the point of view of the subsequent developments in analytic chemistry. Indeed, Antonio Clericuzio argues that the correspondence 'shows that Boyle's preoccupations in his relations with the mechanical philosophers was to safeguard the role of chemistry as a discipline independent from physics'.[28] It is therefore rather to this end that Boyle's speculations about the chemical properties of particles should be understood to have been directed. Boyle's corpuscular chemistry can therefore be understood as one of the early bifurcations that occurs in the transitional stage of the differentiation of science from philosophy, and also of chemistry from physics, which is only retroactively reconnected with the plane of reference characteristic of analytic chemistry, marked by the proper names of Lavoisier and Mendeleev.

One of the differences between science and philosophy that Deleuze and Guattari consider to be 'impossible to overcome' is that 'proper names mark in one case a juxtaposition of reference and in the other a superimposition of layer' (*WIP*, 128). In science proper names play the role of '*partial observers*' that are 'installed like a golem in the system of reference' of 'the things studied' (130), that is, as 'points of view in things themselves' (132). For example, the Boyle of Boyle's Law, which states the inverse proportionality of pressure and volume at a given temperature, is also the Boyle who is retroactively understood to have installed himself by means of his very speculations as a partial observer at the level of the interacting corpuscules in his own experiments on nitre. Whereas for philosophy, proper names play the role of '*conceptual personae*' which differentiate the planes of immanence of different fragmentary concepts that are superimposed on one another in the process of the construction of concepts (Scotus after — because superimposed upon — Descartes). In addition to Boyle's Law, the proper name Boyle would therefore also designate the bifurcations and ruptures of corpuscular chemistry from physics and the mechanical science of the time; and Lavoisier would designate the bifurcations and ruptures of analytic chemistry that are only retroactively reconnected with Boyle's corpuscular chemistry. It is only by virtue of this relation of bifurcation/rupture and retroactive reconnection that the proper names of Boyle and Lavoisier should be understood to be juxtaposed in relation to one another in the ramified time of science.

A closer examination of the correspondence between Boyle and Spinoza is required in order to support the assessment of the distinction between their replies to one another as being characteristic of the distinction between science and philosophy. In Letter VI, Spinoza responds to Oldenburg giving his 'judgment of what [Boyle] has written (. . .), noting certain things which seem to [him] obscure, or inadequately demonstrated'. He begins by offering a clear and concise account of Boyle's experiment on nitre, thereby demonstrating that he understood the point of Boyle's interpretation. He writes that Boyle 'infers from his experiment concerning the reconstitution of Nitre that Nitre is something heterogeneous, consisting of fixed and volatile parts, whose nature (so far as the Phenomena are concerned, at least) is nonetheless very different from the nature of the parts of which it is composed, though it arises solely from the mixture of these parts' (Letter VI).

Boyle considered his experiments on nitre to illustrate that the entire body of nitre could be analysed, or decomposed into its constituent parts — fixed nitre and spirit of nitre — that were different or heterogenous, and that the initial nitre could then be reconstituted by 'redintegration' from these different parts. Boyle inferred that the corpuscules of the constituent parts persist unchanged throughout the reactions, and that the reactions were explicable on the basis of his corpuscular chemistry.[29] The essential point is the process of analysis-synthesis that highlights the reversibility of the transformation that nitre is subject to. The modern chemical equations for the reaction, which were only determinable in relation to the plane of reference of science post Lavoisier/Mendeleev but which will serve to assist in assessing both Spinoza's and Boyle's experiments, are as follows:

$$4KNO_3 + 3C \rightarrow CO_2 + 2NO_2 + N_2 + 2K_2CO_3$$

$$NO_2 + H_2O \rightarrow 2HNO_3$$

$$K_2CO_3 + 2HNO_3 \rightarrow 2KNO_3 + H_2O + CO_2$$ [30]

Boyle put hot coal (C: carbon) into the 'nitre' (KNO_3: saltpetre or potassium nitrate), which was decomposed, leaving 'fixed nitre' (K_2CO_3: potash or potassium carbonate). The nitrogen dioxide (NO_2) gas, which when condensed on glass as vapour, or mixed with water (H_2O), produces spirit of nitre (HNO_3: acqua fortis or nitric acid), was left to escape into the open air. Boyle then added spirit of nitre, that was sourced separately, to the fixed nitre until crystals of nitre were formed. The role of carbon in the experiment was unsuspected by Boyle, as was the fact that it combined with certain elements to compose the nitre, and to form carbon dioxide (CO_2) that was released into the air.[31]

In response to Boyle's experiment on nitre, Spinoza argues that in order 'for this conclusion to be regarded as valid', Boyle should have done a further experiment that would have shown that: 'Spirit of Nitre is not really Nitre and cannot be solidified or crystallized without the aid of the alkaline salt' (Letter VI). Without such an experiment, Spinoza considers the hypothesis of the homogeneity of nitre and its spirit, which he will introduce as an alternative explanation of the experiment, not to have been ruled out.

Spinoza also claims that 'what the Distinguished Author says he observed with the aid of the scale (§9)', does 'nothing to confirm his conclusion' (Letter VI). In §9 of the *Essays*, Boyle reports that he weighed the spirit of nitre necessary to fully dissolve the fixed nitre and compared this with the weight lost by the nitre when it was separated from its spirit, and found that 'the weights were nearly, but not quite equal'.[32] Spinoza maintains that Boyle's quantitative check did not support his case. He argues that Boyle should have at least tried to show that, in the decomposition of nitre, a given quantity of nitre always produces that same quantity of fixed nitre or that the quantity of fixed salt obtained was always proportional to the amount of nitre required to produce it. While Spinoza's reasoning is quite justified in this criticism, it should be noted that he too fails to satisfy these requirements in his own experiment, thought he does acknowledge that he does not have the means to do so. Despite this, he incorrectly claims that, if this quantitative check 'could be made accurately, it would completely confirm what I wished to infer' (Letter VI), that is, the homogeneity of nitre and its spirit.

Spinoza considers this to be 'the simplest explanation of (. . .) the reconstitution of Nitre'. His alternative explanation of the experiments is that the differences observed between nitre and its spirit are due to the different states of motion and rest of the particles: 'the particles of the Nitre are at rest, whereas those of the spirit of Nitre, having been considerably stirred up, keep one another in motion' (Letter VI). The distinct properties of the nitre and of its spirit, which Boyle had used to support the inference that they were heterogenous particles endowed with different natures, are here explained by Spinoza in purely mechanical terms. For Spinoza it is primarily and principally according to differences in movement and rest that material and physical differences must be explained. It is this principle that Spinoza uses to explain not only the observed differences between nitre and its spirit, but also the differences in their taste and inflammability.

As for the fixed salt, Spinoza considers it to be an impurity in the nitre and to do 'nothing to constitute the essence of Nitre' (Letter VI), the analysis of which he considered to be a process of purification. The nitre becomes brittle upon heating, which allows the separation of the particles of nitre that are in motion from its impurities which remain at rest as the fixed nitre. Spinoza is correct to recognize that the fixed nitre contains impurities, however, he too fails to correlate this with the carbon introduced into the process by the coal. Any quantitative check of the process that did not take into account the role of carbon was bound to fail. Both Spinoza and Boyle err in this respect.

Spinoza's mechanical account of the reconstitution of nitre is as follows: 'with the aid of water or air, the fixed salt is loosened and made more flexible, then it is sufficiently able to restrain the impetus of the particles of [A: spirit of] Nitre and to force them to lose the motion they had, and come to rest again (just as a cannonball loses its motion when it hits sand or mud)' (Letter VI).

In response, Oldenburg writes that Boyle 'thinks what you suppose about how it occurs — that you consider the fixed salt of Nitre to be its impurities, and other such things — is said gratuitously and without proof' (Letter XI). Taking up a position with regards to Spinoza's conjectures, Boyle is reported as claiming that spirit of nitre is nitre 'materially', but not 'formally'. Oldenburg presents Boyle's argument as follows: 'materially, indeed, Spirit of Nitre is Nitre, but not formally, since they differ very greatly in their qualities and powers, viz. in taste, smell, volatility, power of dissolving metals, of changing the colors of vegetables, etc.' (Letter XI). Spirit of nitre is 'materially' nitre, since it is a constituent part of it, but it isn't 'formally' nitre, since its nature, as manifested by its properties, is different. For Boyle, fixed nitre is therefore also equally found as such in nitre. But by using this Aristotelian terminology to affirm that nitre and its spirit are formally distinct, Boyle risks problematizing the very distinction that he is trying to establish between his own corpuscular chemistry and the Scholastic doctrine of substantial forms. According to this doctrine, the properties of a natural substance like nitre were determined by its possession of a 'form', which would be destroyed if the substance underwent substantial change such as that brought about in the experiment. The redintegration of nitre was intended to show that such substances could be broken up into more elementary constituents and then made whole again by reuniting the constituents, which persisted unchanged throughout the reactions. So the properties of the whole were to be explained, not by its possession of a form, but by the composition of its parts.[33]

Peter Anstey argues that the use of such anachronistic terminology 'was part of natural philosophy parlance'. So while Boyle explicitly rejects the Scholastic doctrine of substantial forms, 'he is quite prepared to categorize the qualities in Aristotelian terms'. It is also important to note that by Boyle's day the Aristotelian qualities had become 'the *explananda* and not the *explanans*', that is, there was an 'inversion of the explanatory role of the Aristotelian' qualities.[34] So, despite referring to the distinction between nitre and its parts using Aristotelian terms, the actual distinction still required explication, and what Boyle was in fact suggesting is that his corpuscular explanation displaces the Scholastic doctrine in this respect.

In response to Oldenburg's disclaimer that Boyle's 'purpose was not so much to show that this is a truly Philosophic and perfect Analysis of Nitre, as to explain that the common doctrine of Substantial Forms and Qualities, received in the Schools, rests on a weak foundation' (Letter XI), Spinoza writes that 'I did not think, indeed I could not have persuaded myself, that this Most Learned Gentleman had no other object in his Treatise on Nitre than to show the weak foundations of that childish and frivolous doctrine of Substantial Forms and Qualities' (Letter XIII). Spinoza considered this to have 'already been more than adequately demonstrated by Bacon and later by Descartes' (Letter VI). Spinoza is obviously not persuaded that Boyle only proposed to show that the doctrine of substantial forms and qualities has no solid foundation. This opinion is vindicated by Oldenburg's comments that Boyle's aim was 'primarily to show the usefulness of Chemistry for confirming the Mechanical principles of Philosophy, and that he had not found these matters treated so clearly by others' (Letter XI). Nevertheless, this does not detract from the earlier claim that Boyle's aim does seem to have been concerned to displace the Scholastic doctrine of substantial forms. It just affirms that this was not conducted at the level of conceptualization, but rather at the level of experimental demonstration, that is, not philosophically but rather scientifically. So, in Oldenburg's words, Boyle 'has shown that the thing [Analysis of Nitre] occurs thus, but has not discussed how it occurs, which seems to be the subject of your [Spinoza's] conjecture. Nor has he determined anything about it, since that was beyond his purpose' (Letter XI).

In the final letter of the exchange, Oldenburg writes that Boyle 'had only wished to show that the various textures of bodies produce their various differences, that from these proceed quite different effects, and that so long as the resolution to prime matter has not been accomplished, Philosophers and others rightly infer some heterogeneity from this' (Letter XVI). As far as Oldenburg is concerned, he doesn't 'think that there is any fundamental difference between [Spinoza] and Mr. Boyle here' (Letter XVI). Ostensibly, Oldenburg is right: it is possible to distinguish between, on the one hand, ideal corporeal elements, particles or 'the simplest bodies',[35] which are pure rational constructions that are not apprehended in experience, elements whose simplicity renders them in a certain way homogenous with one another; and, on the other hand, real bodies, or modes, as presented in experience with all the differences in complexity of their composition, which could well be interpreted in terms of heterogeneity, on the condition of understanding this heterogeneity to be relative and not absolute. It would be accurate to claim that Boyle did not actually envisage attributing an absolutely heterogenous nature to nitre itself, but wanted only to induce a certain heterogeneity of the complexity of its composition such as was revealed by the experiments that he conducted.[36]

Boyle accounted for the redintegration on the grounds of his speculations about the chemical properties of the corpuscules, and did not make any attempt to deduce them from the mechanical principles, as Spinoza did.[37] Of course, Spinoza's conclusions concerning the redintegration of nitre were false, and as for Boyle, the chemical reaction that nitre undergoes is much more complex than simple decomposition. On the question of determining that all variations of bodies happen according to the laws of mechanics, Spinoza does not think that Boyle's experiments furnish us with a proof more clarifying than other 'readily available experiments' (Letter XIII). This claim is made from the point of view of the supposition of the homogeneity of nitre and its spirit, which requires only a simple mechanical explanation that adds nothing to the already understood principle.

In response, Boyle is presented as claiming 'that there is a great difference between readily available experiments (where we do not know what Nature contributes and what things intervene) and experiments where it is definitely known what things are brought in' (Letter XI). Of course, we have already seen that Boyle commits a number of errors in his judgement about 'what things are brought in' to the experiments. However, despite erroneously supposing fixed nitre to be simply a part of nitre, and therefore not grasping 'the contribution made to his experiment by the coal he used to kindle the nitre', the very distinction that Boyle's speculations posed 'shows the sophistication about experiments that made him a great scientist'.[38] In fact, it is less these particular experiments involving nitre, and more the way in which Boyle systematized, not the corpuscular philosophy, but the practical method for ensuring that the observational and recording process involved in scientific experimentation accumulated into a body of knowledge which was widely disseminated through publication, was accessible to the public, via public demonstrations, and was self-correcting, by means of the requirement to replicate and review the experiments implied in this process. It is this process that marks the parting of ways for science and philosophy, and which constitutes Boyle's lasting contribution to the development of the plane of reference of science. It is this process that is more generally designated by the proper name 'Boyle'.

A number of commentators are also critical of these comments by Spinoza and of what they consider to be Spinoza's understanding of experiments, that is, that he relegates them to the imagination because they deal with the senses, and that he therefore considers them to have no bearing on the principles of reason. This understanding belies the constitutive nature of the imagination in Spinoza's epistemology, which he characterizes as the first kind of knowledge, and its role in the development of reason, or the second kind of knowledge. When, in Letter VI, Spinoza states that fluidity and solidity belong to the class of notions determined by the use of the senses and therefore belong to the imagination, this is not to relegate these notions to some ineffective epistemological category that remains opaque to the understanding, but rather characterizes the kind of understanding that is able to be determined by the senses as being only partial: a partial or limited knowledge. The challenge is to attempt to improve that understanding of these states. The process itself is far from different to that proposed by Boyle's process of experimentation, where a body of knowledge, or plane of reference, is developed in relation to a series of experiments. For Spinoza, a sense of the fluidity of something represents only a partial understanding of the nature of fluidity. The rational component of an understanding of 'the nature of a fluid in general' would be 'to know that we can move our hand [A: in it] in all directions without any resistance, with a motion proportionate to the fluid' (Letter VI), that is, to know that it is the particular state of motion of the particles of the fluid that determines them as a fluid, a knowledge supported by the experiment with your hand. 'This is evident enough to those who attend sufficiently to those Notions which explain Nature as it is in itself, not as it is related to human sense perception. Not that on that account I scorn this history as useless. On the contrary, if this were done concerning each fluid, as accurately and reliably as possible, I would judge it very useful for understanding their special differences. This is something all Philosophers ought greatly to desire, as being very necessary' (Letter VI). The rational idea of a particular fluid, or of the fluidity of the fluid (or solidity of the solid, depending on what the case may be) involves both rational and sensible components. The sense of fluidity is not displaced by the rational idea of fluidity, but remains the sensible component of the rational idea of fluidity and allows the fluidity of different fluids to be differentiated. The sensible component of a rational idea would be a false idea only if it were considered to be all that was necessary in order to understand the idea, that is, if it were itself considered to be the rational idea. The imagination for Spinoza is in this way a constitutive component of reason.[39] Spinoza considers this one observation to indicate 'completely the nature of a fluid' (Letter VI).

Spinoza therefore does not disdain experiments per se; in fact he reproached Boyle in this instance for not having experimented enough. He maintains that Boyle was not consistent enough in his endeavour to give only mechanical explanations of natural phenomena. And it is by placing himself in an experimental point of view, by attempting to replicate the experiments, that Spinoza engages with and is critical of Boyle's experiments. Boyle, as much as Spinoza, is aware of the limited character of the knowledge derived from experiments, even if he does not agree, in the case of these particular experiments, with the placement of this limit. Boyle is concerned to push the boundaries of this limit imposed by mechanics in order to account for his speculations about corpuscular chemistry. He did not want to give just mechanical explanations, but to distinguish properly chemical from physical explanations. Despite the criticisms that each makes of the other's experiments, these criticisms are made from within the same mechanical view of the world. This requires the tacit acceptance by both men of a certain number of general presuppositions associated with the problematic of deploying the mechanical principles in relation to what was in the process of being established as their respective disciplines. All of this occurs prior to the expression of particular points of disagreement in the correspondence.

The difference between Boyle and Spinoza's respective accounts of the experiment can therefore be understood according to the different way that they each respond to the particular problematic about the nature of nitre, that is, whether it is heterogenous or homogenous.

Spinoza relates this problematic to his philosophy and the distinction between the modes and the substance of which they are the affections, while Boyle relates it to his corpuscular chemistry in an attempt to give ground to his speculations about the function of a particle's chemical properties. The debate between the scientist and the philosopher bears exemplary witness to the emerging distinction between the discipline of science and the discourse of philosophy. The fundamental point, however, is that this division emerges solely because it is between a philosopher and a scientist who are already in agreement on the essentials of the principles of mechanics. Their positions are therefore far from being mutually exclusive, as is suggested by the more scientific assessments of the exchange.

Deleuze and Guattari suggest that science and philosophy take different paths, 'because philosophical concepts have events for consistency whereas scientific functions have states of affairs or mixtures for reference: through concepts, philosophy continually extracts a consistent event from the state of affairs (. . .) whereas through functions, science continually actualizes the event in a state of affairs (. . .) that can be referred to' (*WIP*, 126). What I would like to suggest is that this difference between science and philosophy is characteristic of the difference between the way that Boyle and Spinoza engage with the problem of the redintegration of nitre. The state of affairs to which Deleuze and Guattari refer can be characterized in the example of the Boyle-Spinoza correspondence by the particular experiments that are in question. The event that is actualized would be, from the point of view of science, the hypothesis of the redintegration of nitre. Deleuze and Guattari maintain that responding to such a problem, whether from the point of view of science or philosophy, 'does not consist in answering a question but in (. . .) co-adapting (. . .) corresponding elements in the process of being determined' (133). For philosophy, this consists, as we have seen, in co-adapting problematic or fragmentary concepts (of the past); whereas for science this involves choosing 'the good independent variables'—in the present example these would be nitre and its heterogenous parts: spirit of nitre and fixed nitre—'installing the effective partial observer on a particular route'—that is, Boyle installing himself at the level of the interacting corpuscules in the redintegration experiment—'and constructing the best coordinates of an equation or function' (133). Boyle's failure to recognize the role of carbon in the experiments rather limits his efforts in this respect, though the function could be retroactively represented by the chemical equation of the dual reactions presented above.

Deleuze and Guattari further distinguish philosophy from science by maintaining that when an object — for example, nitre composed of heterogenous parts — is 'scientifically constructed by functions, its philosophical concept, which is by no means given in the function, must still be discovered' (*WIP*, 117). Spinoza is interested in trying to determine whether or not a consistent event, and therefore a concept, can be extracted from the state of affairs, characterized by the experiments in question. Deleuze and Guattari maintain that 'The event is actualized or effectuated whenever it is inserted (...) into a state of affairs; but it is *counter-effectuated* whenever it is abstracted from states of affairs so as to isolate its concept' (159). In support of Oldenburg's disclaimers about Boyle's intentions in the *Essays*, Boyle is interested in effectuating the event, the redintegration of nitre, by inserting it into a state of affairs, that is, by performing the experiments. Whereas Spinoza is interested in counter-effectuating the event, that is, in abstracting from the experiments, which he also performed, so as to isolate a concept of the event, a concept of the redintegration of nitre and therefore of its nature.

Another distinction that Deleuze and Guattari consider to mark the divergence between philosophy and science and that characterizes the distinction between Boyle and Spinoza as correspondents is that 'philosophical concepts act no more in the constitution of scientific functions than do functions in the constitution of concepts' (*WIP*, 161). This distinction is further qualified by the claim that it is only 'in their full maturity, and not in the process of their constitution, that concepts and functions necessarily intersect' (161). So, in the case of the exchange between Boyle and Spinoza, because Boyle's corpuscular chemistry was still in the process of development at the time of the correspondence, it could well be argued that his speculations remained as speculations because the scientific functions, the construction of which his speculations contributed to, had not reached their full maturity, and would not do so until the development of the analytic chemistry much later with Lavoisier and Mendeleev. Boyle's interest in effectuating the redintegration of nitre by inserting this hypothesized event into the state of affairs characterized by the experiments in question and Spinoza's attempts to extract a concept of the nature of nitre from them therefore happen at cross purposes. Deleuze and Guattari lament that it is 'unfortunate when scientists do philosophy without really philosophical means or when philosophers do science without real scientific means' (161). Boyle's speculations about the stability, or the non-disruptability of certain characteristic qualities at a chemical level were ahead of his time. They designated ruptures to come in the plane of reference of the emerging science of the day, which was unable to support their representation; and the philosophy that he drew upon only risked further obscuring the value of his speculations that later proved to be so decisive — thus Oldenburg disavows any conceptual and therefore philosophical characterization of Boyle's experiments. Spinoza too was limited by the resources of the predominantly Cartesian science of the day, so that the very reversion to its plane of reference when mobilizing an explanation of his experiments only further obscured the value of Boyle's speculations. Deleuze and Guattari maintain that 'philosophy has a fundamental need for the science that is contemporary with it (. . .), because science constantly intersects with the possibility of concepts' (162). However, this in no way guarantees that a concept will be constructed. Spinoza was unable to counter-effectuate the hypothesized event of redintegration, that is, render the event consistent by extracting a concept from it, and so was unable to isolate a new or different concept of the nature of nitre from the experiments in question.

Rather than succumb to the temptation to loan to Boyle the conceptions of analytic chemistry, and thereby effect a move that lends support to the superficial polarization of the Spinoza-Boyle correspondence as that between a quintessential rationalist and an experimentalist, what I have attempted to demonstrate in this paper is that the rather problematic nature of the exchange provides the focus for an examination of the very divergence that was beginning to emerge between the disciplines of science and philosophy, and indeed encapsulates an example of this very divergence. Despite the fact that their work is similarly grounded on the principles of mechanics, it is the very transitional nature of Boyle's speculations on corpuscular chemistry that provide grounds for distinguishing between the respective projects that they each championed in their correspondence. Boyle's corpuscular chemistry can be understood to be one of the early bifurcations that occurred in the transitional stage of the differentiation of science from philosophy, and of chemistry from physics, which is only able to be retroactively reconnected to the plane of reference characteristic of analytic chemistry. Spinoza took the image of science that Boyle was attempting to construct quite seriously, though in the correspondence he made the mistake of putting forward a simplistic image of it, one that had no scientific value for Boyle. The solutions that Spinoza attempted but was unable to offer to the problem of the redintegration of nitre were philosophical solutions, whereas those that were in nascent form in Boyle's speculations were indeed scientific solutions, and their value to the development of analytic chemistry has been retroactively vindicated.

NOTES

1 Robert Boyle, *Certain Physiological Essays, written at distant times, and on several occasions* (London, Herrington, 1661). The book contained 'A physico-chymical essay, containing an experiment, with some considerations touching the differing parts and redintegration of salt-petre', and was published in Latin the same year as *Tentamina quaedam physiologica, diversis temporibus et occasionibus conscripta* (1661). It is the Latin edition that Spinoza would have read.

2 Gilles Deleuze and Félix Guattari, *What Is Philosophy?*, translated by Graham Burchill and Hugh Tomlinson (London, Verso, 1994). Henceforth *WIP*.

3 See Henri Daudin, 'Spinoza et la science expérimentale, sa discussion de l'expérience de Boyle', *Revue d'Histoire des Sciences* 2 (1948), 179–90. See also A.R. Hall and Marie Boas Hall, 'Philosophy and natural philosophy, Boyle and Spinoza', in *Mélanges Alexandre Koyré*, vol. 2 (Paris, Hermann, 1964).

4 For a defence of Spinoza from the point of view of the inconsistent ontological status of Boyle's elements of matter, see Christopher Lewis, 'Baruch Spinoza. A Critic of Robert Boyle, On Matter', *Dialogue. Journal of Phi Sigma Tau* 27:1 (1984), 11–22; republished in *Spinoza*, edited by Genevieve Lloyd, Critical Assessments of Leading Philosophers, vol. 1 (London, Routledge, 2001). For an examination of the exchange from the point of view of their respective religious speculations, see Luisa Simonutti, 'Spinoza and Boyle: Rational Religion and Natural Philosophy', in *Religion, Reason and Nature in Early Modern Europe*, edited by Robert Crocker (Dordrecht, Kluwer, 2001), (117–38).

5 See Peter Anstey, 'Robert Boyle and the Heuristic Value of Mechanism', *Studies in History and Philosophy of Science* 33 (2002), 164.

6 Peter Anstey, *The Philosophy of Robert Boyle* (London, New York, Routledge, 2000), 4.

7 See Pierre Macherey, 'Spinoza lecteur et critique de Boyle', *Revue du Nord* 77 (1995), 733–74 (744). Macherey's account of the Spinoza-Boyle correspondence is the most balanced to date.

8 Elkhanan Yakira, 'Boyle et Spinoza', *Archives de Philosophie* 51:1 (1988), 107–24 (109). Although Yakira recognizes the limitations of such a schematic representation of the distinction between Boyle and Spinoza, his response recasts this very distinction in the form of a paradox that can be resolved dialectically in Boyle's favour.

9 See Pierre Macherey, 'Spinoza lecteur et critique de Boyle', 769.

10 See Peter Anstey, 'Robert Boyle and the Heuristic Value of Mechanism': 'Boyle repeatedly designates texture as a mechanical affection of matter, even though it is not possessed by atomic corpuscles' (164).

11 Peter Anstey, *The Philosophy of Robert Boyle*, 6.

12 François Zourabichvili, *Le Vocabulaire de Deleuze* (Paris, Editions Ellipses, 2003), 67.

13 Daniel W. Smith, 'Axiomatics and Problematics as Two Modes of Formalisation: Deleuze's Epistemology of Mathematics', in *Virtual Mathematics: the Logic of Difference*, edited by Simon Duffy (Manchester, Clinamen Press, 2006), 181.

14 See François Zourabichvili, *Le Vocabulaire de Deleuze*, 67.

15 For an explication of the terms 'point of condensation' and 'neighbourhood', see Simon Duffy, 'The Mathematics of Deleuze's Differential Logic and Metaphysics,' in *Virtual Mathematics*, 131, 126–32.

16 This example is presented by Ilya Prigogine and Isabelle Stengers in *Entre le temps et l'éternité* (Paris, Fayard, 1988), 162–3. See *WIP*, 225, note 1.

17 Deleuze and Guattari write that: 'The life of philosophers, and what is most external to their work, conforms to the ordinary laws of succession; but their proper names [as conceptual personae] coexist and shine either as luminous points that take us through the components of a concept once more or as the cardinal points of a stratum or layer that continually comes back to us' (*WIP*, 59).

18 For further discussion of this distinction see Simon Duffy, *The Logic of Expression: Quality, Quantity and Intensity in Spinoza, Hegel and Deleuze* (Aldershot, Ashgate, 2006), 253–70.

19 Peter Anstey, *The Philosophy of Robert Boyle*, 4. For further discussion of the heuristic value of mechanical philosophy to Boyle see Anstey, 'Robert Boyle and the Heuristic Value of Mechanism'.

20 See A.R. Hall and Marie Boas Hall, 'Philosophy and natural philosophy, Boyle and Spinoza', 256.

21 Pierre Macherey, 'Spinoza lecteur et critique de Boyle', 762.

22 Peter Anstey, 'Robert Boyle and the Heuristic Value of Mechanism', 173.

23 'Robert Boyle and the Heuristic Value of Mechanism', 164.

24 *Ethics*, II, P13, L1. All quotes from the *Ethics* and the correspondence are taken from Benedict de Spinoza, *The Collected Works of Spinoza Volume I*, edited and translated by Edwin Curley (New Jersey, Princeton University Press, 1985). The Letters are hereafter referred to by their number.

25 Robert Boyle, *The Origin of Forms and Qualities according to the Corpuscular Philosophy* (London, 1666).

26 See Peter Anstey, *The Philosophy of Robert Boyle*, 108.

27 See *Ethics*, II, P13, L3, A3.

28 Antonio Clericuzio, 'A Redefinition of Boyle's Chemistry and Corpuscular Philosophy', *Annals of Science* 47 (1990), 561–89 (577).

29 See Alan Gabbey, 'Spinoza's Natural Science and Methodology', in *The Cambridge Companion to Spinoza*, edited by Don Garrett (Cambridge/New York, Cambridge University Press, 1996), 178.

30 Henri Daudin incorrectly balances the redintegration (synthesis) equation by mistakenly putting a 2 in front of the carbon dioxide. See 'Spinoza et la science expérimentale', 180.

31 The production of carbon dioxide in similar experiments using coal was discovered by Joseph Black in 1754, who described it as 'fixed air'.

32 Curley, *The Collected Works of Spinoza*, 174, note 17.

33 *The Collected Works of Spinoza*, 173, note 15.

34 See Peter Anstey, *The Philosophy of Robert Boyle*, 22.

35 *Ethics*, II, P13, L3.

36 See Pierre Macherey, 'Spinoza lecteur et critique de Boyle', 768.

37 Antonio Clericuzio, 'A Redefinition of Boyle's Chemistry and Corpuscular Philosophy', 577.

38 Curley, *The Collected Works of Spinoza*, 199, note 72.

39 For a discussion of the constitutive role that is played by the imagination in Spinoza's epistemology, see Genevieve Lloyd, *Part of Nature: Self-Knowledge in Spinoza's Ethics* (Ithaca, NY, Cornell University Press, 1994), 63 ff., and the discussion of 'common notions' in Gilles Deleuze, *Spinoza: Practical Philosophy*, translated by Robert Hurley (San Francisco, City Lights Books, 1988), 54–8.

Becoming Interdisciplinary: Making Sense of DeLanda's Reading of Deleuze

David Holdsworth

Abstract

Despite the generally positive reception of Manuel DeLanda's *Intensive Science and Virtual Philosophy* (2002), it has been pointed out that DeLanda's reconstruction of Deleuze's ontology has concentrated almost exclusively on the processes of becoming actual, and has thus far failed to address the processes of becoming virtual. In this article, I suggest a way of reading DeLanda which recovers for mathematical practice a capacity to clarify the meaning of *events* as they arise within a synthetic process of becoming interdisciplinary. First, I attempt to show that modern mathematical practices can be understood as already characterized by a denial of essentialism in just the way required by Deleuze. Second, I argue that mathematical practice, so understood, is of a piece with literary practice, in the sense that both can be understood as the free and transformative production of concepts within the space of what Deleuze has characterized as a form of the comedic. I conclude with an outline of an argument showing that the transformative and the comedic can be embraced within a realist philosophy of the kind so clearly evident in DeLanda's reconstruction of Deleuze's ontology.

Keywords: DeLanda, interdisciplinarity, Deleuzian ontology, mathematics, literary criticism

DeLanda's reconstruction of Deleuze's ontology provides an invaluable means of entry into the Deleuzian text, and the aim here is to explore this reading.[1] I come to this project as a quantum physicist and post-empiricist philosopher of science who has become increasingly influenced by themes from both French and German philosophy. I am actively interested in the possibility of reconciliation between Anglo-American and continental philosophical traditions.[2] I would identify as examples of such projects Richard Bernstein's undertaking in *Beyond Objectivism and Relativism* to trace the common preoccupations with the theme of rationality,[3] and Karl-Otto Apel's project, starting from his reading of C.S. Peirce, to find shared insights across analytical philosophy and German hermeneutics.[4]

DeLanda's reconstruction

DeLanda's reading of Deleuze gives us important insights into the frequent allusions to mathematical concepts within the Deleuzian text. DeLanda's reconstruction makes explicit the formal constructions within the theory of differentiable manifolds that are the basis of Deleuze's notion of a multiplicity. Although Deleuze rarely made these ideas explicit himself, it seems clear that DeLanda has provided an insight into just how the mathematical artifacts within the theory of differentiable manifolds do indeed constitute the ground of Deleuze's 'reflection' on a realist ontology of processes, specifically, a realist ontology of multiplicities.[5] As DeLanda makes clear in the second half of his book, it is not enough to clarify the figures within Deleuze's ontology of multiplicity. It must also be clarified that realism can be supported without recourse to an essentialist account of the very artifacts that serve to 'represent' multiplicities. An account must be provided that does not suppose the presentation of virtualities as pre-existing individuals.

DeLanda's reconstruction has both a mathematical and a methodological aspect. As far as the mathematical aspect is concerned, it will be necessary to radically rethink mathematical practice itself in such a way as to free it from its Platonic resonances, resonances that still echo in many constructivist philosophies of mathematics. And yet, it must be *realist*, understood in its most general philosophical sense, namely that it must acknowledge a world that exists independently of human intervention and understanding, without assuming that everything that exists does so by virtue of belonging to categories, that are characterized by those properties possessed by all and only its members. The version of realism that I defend, and that I consider to be at least consistent with a Deleuzian ontology, is a realism closely aligned with a *normative objectivity*, that is, a notion of objectivity that is fundamentally *methodological*, not ontological, but that makes contact with reality in terms of the *rationality* of our beliefs. The latter, in turn, is to be assessed in terms of our *critical commitment* to notions of *invariance*. (These notions will all be given further clarification below.)

As for the methodological aspect, the epistemology of problems that Deleuze develops must be shown to be of a piece with the ontology of multiplicities, in such a way that the objectivity of science, understood as a methodological principle, reinforces the realist dispositions within the ontological interpretation of multiplicities as non-individuated realities. This involves a radical re-understanding of scientific practices, a re-understanding that respects the empirical analysis of productive causes. Moreover, it must also re-theorize the productive effects of the active intervention and transformative possibilities that accompany scientific activity. In short, we must theorize methods of inquiry that are immanently within the discursive space of that inquiry, but we must do so in such a way as to be able to make judgements about that which exists outside the space, thus making realism feasible for theoretical practices that operate immanently within their discursive spaces.

With respect to the methodological aspect, my objective is to defend a strong notion of interdisciplinarity consistent with the Deleuzian ideal of *scientist becoming-philosopher* and *philosopher becoming-scientist.* In this context I propose to take seriously Deleuze's notion of humour, and in this way, to open up, at least tentatively, a way of recognizing literary production as potentially continuous with scientific production. With respect to the mathematical aspect, to which I turn first, I shall focus on the possibility of a philosophy of mathematics according to which the objects of mathematics are always already *simulacra*, neither copies of some ideal original, nor arbitrary constructions of an undisciplined imagination. The *methodological realism* that emerges is at least consistent with Deleuze's philosophy insofar as it takes into account notions of the active transformation of human agency, and the active agency of the world. In this context I propose to take seriously modes of interdisciplinarity found outside science, in such areas as literary criticism, in order to better understand the processes of *becoming-interdisciplinary*.

The discussion of mathematical practice, as it arises in particular within the mathematical natural sciences, will leave us with a methodological problematic concerning the appropriation of theories within particular domains of empirical inquiry. I argue that this problem is not different from the parallel problem as it arises within other forms of intellectual production, including literary production, that in both cases there is an irreducible requirement for judgement, that this judgement is in fact a form of political judgement, and that this view can be consistently held alongside the account of normative realism that will be elaborated below.

Mathematics as an Immanent Practice

In attempting to reconstruct Deleuze's ontology by means of an explicit interpretation of constructions within the theory of differentiable manifolds, DeLanda faces a classic dilemma, one often faced by those who work in the foundations of logic. The typical dilemma takes the following general form: we want to give an *explication* of a concept or an idea, or alternatively, show that the concept or the idea is *avoidable*, in situations where it appears to be impossible not to take the concept or the idea for granted. An excellent example of this arises in programmes to establish non-standard logics, such as quantum logic. In this case it is frequently pointed out that the arguments in favour of the non-standard logic are articulated according to the canons of classical logic, thus undermining the very thesis being defended. Thus, the quest for an explication fails due to circularity, or the quest for a problematization fails because we are irreducibly dependent upon the concept or the idea to articulate apparent alternatives. So the apparent alternatives fail since they are reducible to the original concept or idea, a concept or idea that thus seems to be ineliminable. In the case of DeLanda's reconstruction, this problem arises because the attempted explication of multiplicities comes across as a *model*. DeLanda is well aware of this and expends a great deal of energy throughout his book in order to show that the entities being modelled are not themselves eternal structures with essences: essences captured by the modelling or by the axioms that found the model in question. The problem can be seen in a particularly clear form when we consider that the models are invariably constructed as *sets*. We are interested in spaces, and these are represented as sets of points. But points are the ultimate individual. Indeed, set theory itself can be thought of as the theory of radically individuated entities. The thing that distinguishes two elements of a set is simply that they are not identical to each other.

The theory of differentiable manifolds was developed in the nineteenth century, consisting originally in Gauss's demonstration that curves and surfaces could be construed as spaces. This idea is part of the popular interpretation of the theory of relativity (Albert Einstein's theory of gravity) and the conceptual innovation that it carried, requiring us to think of space as curved. Thus, in popular accounts of relativity theory, we are asked to imagine that the surface of a sphere is an example of a curved space, truly a space, that is finite but unbounded (having no identifiable boundary), and that is 'constructed' by imagining a flat surface being bent into a third dimension. Or more simply still, we are asked to imagine a circle, constructed by bending a line into a second dimension. But what is it to say that these are spaces, as opposed to figures that can be drawn inside the more familiar geometrical spaces in which bodies move? For Gauss, to say that these curved surfaces are spaces is to say that their properties can be characterized in terms of local differential operations, internal (or immanent) to the space in question. Roughly speaking, they can be axiomatized[6] simply by dropping Euclid's last postulate (the postulate of the parallels).[7]

The significance of this for Deleuze, as DeLanda emphasizes, is the idea of geometric immanence, the idea that Gauss taught us how to conceive of curved spaces without having to appeal to a higher-dimensional embedding space that is transcendent with respect to the space of interest. And indeed, metaphorically, this idea generalizes to discursive spaces, that is, it generalizes to the notion that our discourses are confined to the closed spaces generated by their codes,[8] or their rules of formation.[9] My project here can be understood as an attempt to show how discrete disciplinary practices, as diverse as mathematical physics and literary criticism, share an interdisciplinary capacity to theorize the outside. The strategy here is to begin by interpreting the Deleuzian appropriation of differential geometry, according to DeLanda's reading, as a particularly clear example of this procedure at the level of mathematical practice.

It is important to keep in mind here that the work of Gauss (and later Riemann, who generalized Gauss's work) would not have been possible if it were not for the development of analytic or Cartesian geometry. In analytic geometry spaces are represented by sets of points, but more specifically, sets of points understood as ordered sets of numbers. So coordinate systems enable us to give points names. A point in a plane is represented by the pair of real numbers (x,y), where x is its displacement along one axis, and y is its displacement along the other. In a Cartesian coordinate system, the axes are at right angles to each other. The upshot is that geometry can now be done analytically, since geometric theorems can be established algebraically. Here, 'algebraically' means that the relevant equations are simple equations of the kind we learn to solve in school. It will have a more abstract meaning as we proceed. Prior to the invention by Descartes of analytic geometry, geometry had been synthetic: that is to say it was based upon the operations that we can perform with ruler and compass. Synthetic geometry had quickly become axiomatic with Euclid, who was able to establish the principles of geometry with his five postulates, including the axiom of the parallels.

The very recent development of what has been called synthetic differential geometry contributes to our understanding of Deleuze. As we saw above, DeLanda's greatest challenge in offering a formal reconstruction of Deleuze's multiplicities, is to show that we can go down this road while avoiding the charge that multiplicities, under this reconstruction, are themselves timeless essences, which would of course undermine Deleuze's project to develop a pure theory of becoming that is radically free of the notion of essences. DeLanda undertakes to defend this claim in terms of a sophisticated series of philosophical transformations, each intended to be offered in a Deleuzian spirit, in accordance with the epistemology of problems characteristic of Deleuze's philosophy. The aim here is to reinforce this strategy, without undertaking to evaluate its general success, by adding an argument to show that, even at the level of mathematical practice, we can justify the anti-essentialist claim. It will also be argued that this approach is consistent with a version of methodological realism.

Axiomatic systems in contemporary mathematics do not simply involve the development of a set of postulates that, taken together, are sufficient to draw all the valid inferences about some domain of interest. While it is the case that in practice mathematical axiomatic theories typically arise out of some domain of interest, the mathematical theory itself is understood model-theoretically. Roughly, this means that the theory is taken as a theory of anything that happens to satisfy the axioms. This gives rise to a certain kind of tautologousness, in the sense that it is taken for granted that, unless the domain does satisfy the axioms, then it is simply not a target of our discourse. The originating model may have been a particular domain of interest, but the resultant mathematical theory is about that domain only insofar as it is, as intended, among the models for the theory. But the theory is about all models, and its theorems are construed as statements that are necessary in the sense of being true in any model for the theory.

Alternatively, we might want to say that what is essential about a particular kind of thing is that which is invariant across all the theories that pick out some feature of the system. So, according to this view, every system might be such that, for every property that applies to it, that property is shared by many other systems. But what is *essential* about the system in question are the core properties, the complete set of which is possessed by no other system. This is one possible view to take, but it would be extremely difficult to imagine a method for identifying these properties. Unlike the Platonic form, which we arrive at by contemplation on the ideal limit of the concept, the view just sketched would require a much more robust kind of imagination, capable of scanning all of the concepts that might be applicable to it and all of the other systems, for each concept, to which that concept might apply. In other words, a kind of infinite empirical imagination would be required.

The argument here is that the model-theoretic axiomatic turn seems best suited to a strictly empirical point of view, in the sense that, although the notion of a model admits arbitrarily many alternative systems, it remains the case that the intended model is at the centre of our attention. The theory gives us an instrument for discovering a variety of properties of the system, which must be true of it as long as the axioms are true of it, and at the same time, provides multiple insights into other systems that also satisfy the axioms. However, neither the system first contemplated nor any other model need be imagined as the original, of which each model is a copy. The theory of groups is about anything that is a group but does not require the existence of an original or prototype group to serve as the ultimate ontological target of the theory.[10]

Mathematics as a synthetic practice

Recent developments in algebraic logic and algebraic geometry suggest the possibility of a synthetic differential geometry.[11] The general features of this new approach will be outlined here, without presuming to clarify in what sense differential geometry as such is recovered within the new framework. However, if the interpretation of this new foundation for mathematics that I shall defend can be rigorously established, it goes a long way towards establishing that mathematics, as a practice, and as a body of synthetic knowledge about its objects of interest, can be conducted constructively without relying on an irreducible (essentialist) concept of a set of elements. If this is right, then we have gone a long way towards establishing a Deleuzian philosophy of mathematical ontology at the level of mathematical practice itself.

The approach pioneered by William Lawvere, working in the area of algebraic logic, undertakes to show that we can replace set theory as a foundation for mathematics by axiomatizing particular kinds of mathematical categories in a way that is independent of the concept of a set. The language here can be misleading, since the concept of a mathematical category is reminiscent of the philosophical categories, such as those invoked by Kant, as categories of the understanding within a transcendental philosophy. Clearly, this does not sound congenial to a Deleuzian ontology of multiplicities, theorized as radically non-essentialist. Indeed, the intuitive idea behind a mathematical category, as first conceptualized by Eilenberg and MacLane,[12] involves the notion of a collection of mathematical objects of the same type, along with the structure-preserving morphisms that map between the objects. These sound just like sets, and indeed, Eilenberg/MacLane category theory can be presented as a branch of classical algebra. But this notion of algebra is more general than the notion used earlier when discussing analytic geometry. An algebra is any mathematical system of elements, along with operations upon the elements, such as addition, and a principle of closure, according to which, if A and B are in the algebra, so is $A + B$. Clearly, this way of talking is still dependent on the notion of a set of radically individuated elements. What Lawvere was able to do, in two steps that took him from categories in general to very specific types of categories, which he called 'topoi', was to theorize axiomatically a generalized concept of a set, strictly in terms of the morphisms (or structure-preserving maps) that is not characterized in terms of radically individuated elements, but rather, in terms of universal mapping properties. The claim made here in the context of a Deleuzian ontology of virtualities and epistemology of problems, is that Lawvere's approach makes it possible to claim coherently that mathematical problematizations, such as the problematization of local differentiation addressed by Gauss, can lead to a practice of theorizing abstract structures without assuming that models are sets. Indeed, set theory is recovered as just one example of a topos, among others, including the particular topoi that make a detailed synthetic treatment of differential geometry possible. But the view of mathematical practice that emerges from this is pragmatic, even transformative, one that was characterized by its founders as intuitionistic, in the sense of intuitionism as a foundation for mathematics.

At this point it is helpful to invoke DeLanda's articulation of a formal definition of a multiplicity: '*A multiplicity is a nested set of vector fields related to each other by symmetry-breaking bifurcations, together with the distributions of attractors that define each of its embedded levels.*'[13] The key point is that Deleuze's concept of a multiplicity arises out of the formal mathematics of differentiable manifolds, being multiplicities in the sense that they are multiples of physical realizations of systems that exhibit the same dynamical processes of internal transformations. Something similar is going on here to what is described above for model theory in general. A theory admits a variety of models. The theory is not a theory of any one model in particular, but theorizes an aspect of anything that happens to be a model for the theory. Deleuzian multiplicities, in accordance with DeLanda's definition, are not formally equivalent to models, but the notion of multiple physical systems exhibiting the same structure of nested cascades of symmetry-breaking transformations is a cognate concept.

Methodological Realism

This discussion of mathematical practice, and the interpretation of abstract models, has gone part way towards an account of simulacra, substantiating the claim that models are not copies of originals in a Platonic sense, but not yet justifying the claim that they are not simply arbitrary constructions. The claim that I want to defend, a claim that is strongly consistent with Deleuze's philosophy of transformative production, is that the realism that can be defended arises out of a notion of *methodological objectivity*, *active transformation* as a feature of scientific production, and a notion of *invariance*. The problem with notions of invariance for a reading of Deleuze, at least in relation to the tradition that sees invariance as a criterion of objectivity, is that we normally think of it as invariance across a changing dynamical variable, such as time, or location. Persistence across variations in spatial perspective, or simply persistence across time, are simple examples of invariances that constitute *prima facie* evidence for the objective reality of something that underlies our perceptions. The problem is that, within a philosophy of pure difference, this traditional attitude towards invariance seems to break down, and we must work out a notion of invariance, as a criterion of objectivity, that makes sense within a Deleuzian framework of difference and repetition. The solution to this problem, as DeLanda makes quite clear in general terms, is to emphasize notions of active transformation and to work out ways of recognizing that which is invariant with respect to these sorts of operations.

We arrived, somewhat ironically, at a notion of the model as simulacrum (in its first sense, as a model that is not a copy of any original) by moving to ever-increasing levels of abstraction, only to come to an understanding of models that is radically suspended within an active practice of mathematical creativity. Similarly, we can arrive at an understanding of invariance in transformative terms by passing through ever-increasing levels of abstraction, from invariance across time and space, as discussed briefly above, to invariances across formal languages, to invariances across cultures, and some version of inter-subjectivity as a form of invariance. The latter form of invariance is of course worrisome at its most general level, because subjects are themselves the products of the play of power, including the play of power that takes place between rival theories within scientific or literary production. However, putting that aspect aside, at least temporarily, there are forms of invariance across languages that arise already at the level of mathematical discourses, which are very informative. On the face of things, the theory of partially-ordered sets (sets with a partial order defined upon them), such that for at least some pairs of elements, A and B say, an ordering applies to that pair, written $A > B$. The important point to make here, is that the formal languages used to articulate these theories are radically different, the one being first-order, the other being second-order. And yet, they are equivalent in a significant sense that can be proven within category theory. From the perspective of the discussion here, this means that everything that is a model for one theory, is a model for the other, and yet theorizing things the one way is a radically different *conceptualization* of them from theorizing them the other way.

One can ask, and this becomes even more interesting if we ask the question, or pose the problem, at the level of physical systems, what are the invariant properties across these models, picked out as invariants across these radically different languages? In a sense, having done so, we can discard the languages that were used to theorize these objects, staying focused on the invariants that, through the discourses generated by these languages, we can be said to have discovered. If we take the view, first articulated by John Winnie in the context of his work on the special theory of relativity, that objectivity is equivalent to a notion of theoretical determinateness,[14] that the objective features of models are the features that are invariant across radical transformations across languages, we arrive at the following point of view about objectivity and notions of the real: notions of objectivity arise meaningfully in science only at the level of practice, but practices are characterized by the dynamic role of theories within those practices. So for anyone, any practitioner of any practice within any community of thinkers/speakers, to address questions of objectivity, questions must be asked, and cognate problems posed, relative to the theories that are tentatively held. We must ask: 'What is it rational for me to take to be the objective or theoretically determinate features of the world, in view of my commitment to a particular theory?' Put differently: 'What are the theoretically determinate features of any world that is a model for my theory?' This question admits a precise answer within the resources of a particular theoretical practice. For example, it turns out that the theoretically determinate or objective features of any world satisfying the quantum theory are not the first-order properties of things, as is the case for classical mechanics, where they can be construed as the essential properties of things, but rather certain second-order relations, that formally are what represent states (but not 'states of affairs').[15] The first-order properties of things are indeterminate in the sense of this approach, and thus cannot be construed as essential.

What emerges here, again, is a picture of science as an active practice, a practice that constantly includes the checking of beliefs against experiences of invariance, and checking theories for reliability in the face of transformative practices that seek alternate formulations of possibly equivalent theories, as well as the production of alternate theories, that open up new horizons of problems and questions. The upshot, again somewhat ironically, since we arrived in this place through a process of aggressive abstraction with respect to language, is that we come to a view of scientific practice that incorporates abstract theories into a pragmatic process that produces theories actively and interprets their models as simulacra, simulacra because they are not copies of any original, but now understood as having pragmatic, transformative significance vis-à-vis an active empirical process of becoming-critical, inventing new theories and new concepts as old ones fail, but respecting the agency of nature at the same time. It is in this sense that what we arrive at is a *methodological realism* that valorizes the free production of theories and concepts, but does not collapse into a pure constructivism.

This way of incorporating the active agency of the theorist into inquiry makes clear that there are significant methodological similarities between scientific and literary production. These are not superficial similarities, but rather robust similarities that take us to the heart of interdisciplinarity, the third term, where we encounter the becoming-philosopher of the scientist and the becoming-scientist of the philosopher. It might be objected here that philosophy is misrepresented by eliding it with literary production, and indeed, it is important to come to terms with that relationship, as Richard Rorty undertakes to do in much of his writing.[16] However, this subject merits a full-length discussion in its own right. The final section of this article will continue to focus on the history of theories and the history of concepts, and will move from there into a discussion of

Richard Levin's criticism of 'the new interdisciplinarity' in literary criticism,[17] where it will be argued that his criticism overshoots its target by underestimating the importance of theoretical intervention as a form of political agency.

Becoming interdisciplinary

The scheme outlined here arose as a concern that nothing was being said, by way of motivating the notion of objectivity as theoretical determinateness, about how we come to hold theories in the first place. The philosophy of science has long been concerned with this problematic, as evidenced by the widespread post-empiricist interest in Thomas Kuhn's work on scientific revolutions,[18] and Karl Popper's parallel, but in important ways, opposing, accounts of scientific rationality in relation to the refutation of theories.[19] These sociological and historical turns within the philosophy of science signal an emerging awareness of the dynamics of criticism, and have drawn the focus away from the post-empiricist framework in which everything is still about theories. Indeed, the historiography of science remains focused upon theory-change. Within this tradition, the linguistic turn remains, for the most part, a turn to formal language, to notions of invariance of the sort that have been outlined above, and to transformation at the macro-level of theory-change. There is, however, a whole other tradition that post-empiricist philosophy of science has scarcely noticed, notwithstanding the growth in prestige of continental philosophy in some North American quarters. The linguistic turn, understood as a turn to discourse, the pragmatics of language, and the role these play within the dynamics of power, are scantly acknowledged within post-empiricism, despite its recognition of the historical and social context of practice. Deleuze, like Michel Foucault, was a student of Georges Canguilhem, and shared with him an abiding concern with concepts. Canguilhem, whose work was focused on the history of medicine, believed that biology and medicine did not conform well to the model of the historiography of science routinely brought to bear on what Deleuze would call the 'royal sciences', the grand theories of physics and other mathematical sciences, where history seemed to be a history of theories, and not a history of concepts.[20]

The key area of problematization arises from the question of how to understand the processes by which theories are appropriated, applied, and evaluated, and how institutions can be conceptualized that promote the processes that make normative objectivity possible, while respecting the dynamics of power. Even if it is the case, as argued here, that mathematical and scientific practices demonstrate already, even in their most abstract moments, that models function as simulacra, and if DeLanda is right to define multiplicities in terms of the resources of the theory of differentiable manifolds as nested sets of vector fields, etc., this still does not do justice to the transformational dynamics that Deleuze, especially in his work with Guattari, undertook as an articulation of becoming virtual.[21]

However, the aim here is more modest, namely to offer a partial account of the synthetic process of becoming-interdisciplinary and to make sense of DeLanda's reading of Deleuze as an account of mathematical practice that is already sympathetic to the notion of the pure event. It is my suspicion now that the completion of such a project would be a historiography of science in general (not just biology) that, in the spirit of Canguilhem, would seek to do for a history of concepts what Kuhn and others partially achieved for a history of theories. Such a historiography would have to account for the processes through which we become interdisciplinary by appropriating concepts, transforming them, and bringing them to bear upon problems for which they were originally not intended, a process by which, as Constantin Boundas has put it, 'we find a third term, in between the two, that would facilitate the "becoming-music" of philosophy, and the "becoming-philosophy" of music'.[22] As I have emphasized, such an account would facilitate our understanding of the becoming-science of philosophy, and the becoming-philosophy of science. This middle term, between the philosophical and the scientific disciplines, is the holy grail of interdisciplinarity — becoming-interdisciplinary. A transformative principle of becoming cannot be content with the mere appropriation of concepts, despite my choice of that way of speaking above.

In his critique of interdisciplinarity in literary criticism, Richard Levin quotes Pierre Macherey as follows: 'A rigorous knowledge must beware of all forms of empiricism, for the objects of any rational investigation have no prior existence but are thought into being.'[23] He offers us this quotation at the head of a section on 'aprioritizing theory', just after developing a detailed critique of the methods of literary criticism, consisting of the appropriation of ideas from other disciplines, such as Marxism from political economy, and applying it to the analysis of a text. On Levin's account of becoming interdisciplinary (the Deleuzian hyphen is deliberately omitted here) in literary criticism, there are three steps: selecting a theory, changing the theory (if necessary), and applying the theory. The upshot of Levin's account is a bleak picture of this form of interdisciplinarity, according to which no application of the theory is ever unsuccessful. To put it more bluntly, critics tinker with the theory, Levin argues, in order to make it fit the text. Levin's characterization of this practice is hardly flattering, but his general point is well-taken. Treating theories as *a priori* structures is methodologically worrisome, even in literary criticism, where the worlds being analysed are themselves already the free creation of someone's imagination. Indeed, this is worrisome in any context in which we seek to transfer concepts from one domain to another for the purpose of either critical or empirical assessment. In literary criticism, if the text being criticized is meant as a commentary on possible modes of existence, then the limitations of Marxist theory or Freudian psychoanalysis are also limitations on their usefulness for interpreting a text, even if the gesture of the critic is meant as a political act. For Levin, it is the political aspect of these strategies that problematizes them. Politics requires judgement, and if a becoming-interdisciplinary entails a becoming-political, then it calls for judgement, with all its attendant forms of mediation and making of distinctions.

According to Deleuze, Kant's theory of judgement in the *Third Critique* is already a synthetic resolution of tensions between the *First* and *Second Critiques*, that Kant himself had recognized and undertook to solve in *The Critique of Judgement*.[24] But if judgement is the *sine qua non* of politics, as well as aesthetic theory,[25] and if the *Third Critique* can be read as a kind of political theory, then the resolution is a kind of political resolution. The principal move that Levin makes against the new interdisciplinarity in literary criticism is that it is political, that the practitioners of the new interdisciplinarity are uncritically motivated by political agendas. Accordingly, they are inclined to introduce arbitrary adjustments of the theories being appropriated in order to make the theory conform to the agenda of the critic. An example is tinkering with Freudian psychoanalysis in such a way that self-formations of the female child are pushed back to an earlier moment of development in just such a way as to ensure that Freudian penis-envy becomes, instead, a desire for autonomy from the mother, not the discovery of something lacking. This, Levin argues, is frequently done by some feminist critics, and is politically motivated in a frivolous sense, in order to arrive at a version of Freudian theory that conforms to the interpretation of the text in question in accordance with the political objectives of the critic, to read the text as a narrative of emancipation. My objection to this aspect of Levin's argument is that the validation of theories that he requires is an empirical validation that conforms to the standards of the discipline from which the theory has been appropriated, and this seems entirely unreasonable in the absence of any reason to think that the originating theory is an autonomous practice whose validity can be assessed entirely in terms of its own internal standards. On the other hand, the appropriation of theories in a way that subjects them to arbitrary adjustment in accordance with the standards and expectations of the appropriating practice, such as literary criticism, seems equally problematic.

Theories are appropriated, adjustments are made within them, new concepts are annexed to them, and new contexts of interpretation are created, at least in a practice, that is a becoming-interdisciplinary in a robust sense, not by political fiat, but in a sincere attempt to bring the theory, as a framework of assessment and analysis, into conformity with a political problematic. And as has been argued here, this is as much the case in the mathematical/physical sciences as it is in literary criticism. It is not sufficient to invoke the counter-claim that literature is constituted as the creation of fictional worlds, since it is conspicuously the case that at least some literature is itself a kind of theoretical intervention into our understanding of the world. Shakespeare is full of insights into the human condition, Dickens is full of insights into the social conditions of Victorian London, and Proust is nothing if not a continued 'reflection' on the nature of time and memory. Intervening into Shakespeare or Dickens or Proust through the resources of psychoanalysis is not an exercise in aprioritization as long as the critic continues to be aware of the critical process as an interaction. It is in this sense that Levin overshoots his target, although his observation about the political nature of these interventions is correct, because all interventions, in the sciences and in the arts, are political gestures, requiring judgement in order to be made, and equally requiring judgement in order to be evaluated.

Levin says at the beginning of his paper: 'For Plato all true knowledge must be interdisciplinary, since any inquiry within the confines of a single science has not reached the highest level of the divided line where the separate sciences are *integrated dialectically*' (emphasis added).[26] Deleuze, on the other hand, wants to leave Plato behind, to leave behind the ascent to the Platonic form, to replace the historical figures of irony from Socrates to Kant, with humour. In *The Logic of Sense* he writes:

> The tragic and the ironic give way to a new value, that of humor. For if irony is the co-extensiveness of being with the individual, humor is the co-extensiveness of sense with nonsense. Humor is the art of the surfaces and of the doubles, of nomad singularities and of the always *displaced aleatory point*; it is the art of the static genesis, the savoir-faire of the pure event, and the 'fourth person singular' — with every signification, denotation, and manifestation suspended, all height and depth abolished.[27]

I have emphasized that for Plato the irreducibility of interdisciplinarity arises out of a kind of dialectic, a dialectic across art and science that achieves, at its limit, at the highest level of the Divided Line, an intimacy with the forms. For Deleuze, read through DeLanda, the displaced aleatory point (the contingency of the event) occurs at the surface (of thought and experience), where it is suspended within simulacra, and is irreducibly mediated by judgement. I have tried to show in this article that this is equally the case across disciplines as widely separated as mathematical physics and literary criticism, and tried to establish that all such practices flourish through the free appropriation of theories from other practices, combined with the transformative and productive revision of theories in contexts where new concepts are at play. I have tried to show, in sympathy with Deleuze, and supported by DeLanda's reconstruction, that mathematical practice itself exhibits a kind of immanence internal to its own methods, that mathematical models enter into science as a kind of simulacra, but that a form of normative objectivity is possible within a scheme of methodological realism that is common to all practices that aspire to be critical inquiry. In the end, I have tentatively accepted Deleuze's characterization of this as a form of humour, not so much because I share his conviction with respect to a logic of sense, about paradox, but because some notion of the comedic seems called for if we are to come to terms with the irreducible play among the contingency of theorizing, the judgement implied by our agency, and the constraints of the real.

NOTES

1 Manuel DeLanda, *Intensive Science and Virtual Philosophy* (London, Continuum, 2002).

2 I speak here, perhaps incautiously, of reconciliation. It is not a quest for a fusion, so much as a reaction against the fission of perspectives. I am attracted to the notion of seeking that third place where each becomes the other, i.e., in a strong sense of interdisciplinarity.

3 Richard Bernstein, *Beyond Objectivism and Relativism: Science, Hermeneutics and Praxis* (Philadelphia, University of Pennsylvania Press, 1985).

4 Karl-Otto Apel, *Towards a Transformation of Philosophy* (Boston, Routledge and Kegan-Paul, 1973).

5 Once again, this may be an incautious formulation, since, according to Deleuze, a philosopher does not reflect. The aim here is to establish the practice of using 'reflect' and 'represent' in contexts where artifacts from pre-existing theories, such as those arising in the theory of differentiable manifolds, are being invoked, always keeping in mind that they do not literally represent the philosophical concept that Deleuze is actively introducing.

6 Caution must be exercised here with the notion of an axiom. The particular mathematical theories in question were not historically developed as axiomatic theories, but were developed very much in the spirit of a mathematical construction. There is an ironic twist to the form of my argument, as I shall articulate it, to the effect that it is the path through formal axiomatization that will justify the claim that models are always already simulacra.

7 For every line l and for every point P that does not lie on l there exists a unique line m through P that is parallel to l.

8 See Niklas Luhmann, *Ecological Communication* (Chicago, University of Chicago Press, 1989).

9 See Michel Foucault, 'The Order of Discourse', in *Untying the Text: A Post-Structuralist Reader* (Routledge, London, 1981), 48–78.

10 It needs to be made clear that the intended model is an original in only a contingent sense. It is the conception of that model that motivates the articulation of the mathematical theory. But metaphysically it enjoys no such privilege. There is no metaphysical original.

11 See William Lawvere, 'An Elementary Theory of the category of Sets', *Proceedings of the National Academy of Science of the USA* 52 (1964), 1506–11; republished in *Reprints in Theory and Applications of Categories* 11 (2005), 1–35.

12 Samuel Eilenberg and Sanders MacLane, 'General Theory of Natural Equivalence', *Transactions of the American Mathematical Society* 58 (1945), 231–94.

13 DeLanda, *Intensive Science and Virtual Philosophy*, 32.

14 John Winnie, 'Objectivity and Theory Equivalence', *NSF* Research Proposal, access by permission (1977).

15 David Holdsworth, *A Functorial Semantics for Quantum Logic*, University of Western Ontario, PhD Dissertation.

16 Richard Rorty, 'Deconstruction and Circumvention', *Critical Inquiry* 11 (September 1984), 1–23. Reprinted in Richard Rorty, *Essays on Heidegger*

and Others: Philsophical Papers Volume 2 (Cambridge, Cambridge University Press, 1991), 85–106.

17 Richard Levin, 'The New Interdisciplinarity in Literary Criticism', in *After Poststructuralism: Interdisciplinarity and Literary Theory* (Evanston, Illinois, Northwestern University Press, 1991), 13–43.

18 Thomas S. Kuhn, *The Structure of Scientific Revolutions*, second edition (Chicago, University of Chicago Press, 1970).

19 Karl Popper, *Conjectures and Refutations: The Growth of Scientific Knowledge* (New York, Harper and Row, 1963).

20 Georges Canguilhem, *The Normal and the Pathological*, translated by Carolyn R. Fawcett (New York, Zone Books, 1979). See particularly the introduction by Michel Foucault.

21 This point is made clearly by William Bogard in his review of DeLanda's *Intensive Science and Virtual Philosophy* ('Book Review: How the Virtual Emerges from the Actual', *International Journal of Baudrillard Studies* 2:1 (January 2005), http://www.ubishops.ca/baudrillardstudies/vol2_1/bogard.htm. Accessed on 20/04/05).

22 Constantin V. Boundas, 'Editor's Introduction', in *The Deleuze Reader*, edited by Constantin V. Boundas (New York/Oxford, Columbia University Press, 1993), 2.

23 Quoted in Levin, 'The New Interdisciplinarity in Literary Criticism', 21.

24 See Gilles Deleuze, *Kant's Critical Philosophy: The Doctrine of the Faculties*, translated by Hugh Tomlinson and Barbara Habberjam (Minneapolis, University of Minnesota Press, 1985).

25 See Ronald Beiner, *Political Judgement* (Chicago, University of Chicago Press, 1983).

26 Levin, 'The New Interdisciplinarity in Literary Criticism', 13.

27 Gilles Deleuze, *The Logic of Sense* (New York, Columbia University Press, 1990), 141 (emphasis added).

Notes on Contributors

Simon Duffy is Lecturer in Philosophy and a Postdoctoral Research Fellow in the Centre for the History of European Discourses at the University of Queensland in Brisbane, Australia. He has published a number of articles on the work of Gilles Deleuze, and he has translated a number of Deleuze's Seminars on Spinoza. He is the author of *The Logic of Expression: Quality, Quantity and Intensity in Spinoza, Hegel and Deleuze* (Ashgate, 2006) and editor of the collection *Virtual Mathematics: the Logic of Difference* (Clinamen, 2006). His current research interests include the relation between the work of Deleuze and that of key figures in both the history of mathematics, and the history of German philosophy, particularly Kant and Hegel.

David Holdsworth is Associate Professor of Environmental Studies, and member of the Graduate Program in Theory, Culture, and Politics, at Trent University, Peterborough, Ontario, Canada. He is a theoretical physicist and philosopher of science who now works within the orbit of French and German theory on issues concerning the political and cultural context of scientific practice.

Matthew Kearnes is based in the Department of Sociology, Lancaster University. His research examines the intersections between science, technology and materiality. He is particularly interested in emerging fields of technical development, such as nanotechnology and has published several papers exploring nanoscale research. In this current research on the material innovations of nanotechnology, he has sought to re-think notions of physicality toward an active account of material agency, or 'what things can do'.

John Marks is Reader in Critical Theory at Nottingham Trent University. He has published a number of articles on various aspects of French culture, French philosophy and social theory, and European literature. He is the author of *Gilles Deleuze: Vitalism and Multiplicity* (Pluto, 1998), and co-editor, with Ian Buchanan, of *Deleuze and Literature* (EUP, 2000). His current research interests include debates on cloning and the 'posthuman', as well as the cultural and philosophical reception of molecular biology in France.

Arkady Plotnitsky is Professor of English and Director of the Theory and Cultural Studies Program at Purdue University. He has written extensively on critical and cultural theory, continental philosophy, British and European Romanticism, and the relationships between literature, philosophy, and science. His most recent books include *The Knowable and the Unknowable: Modern Science, Nonclassical Thought, the 'Two Cultures'* (University of Michigan Press, 2002) and, in the series 'Fundamental Theories in Physics', *Reading Bohr: Physics and Philosophy* (Springer Scientific Publishers, 2006, forthcoming).

John Protevi is Associate Professor of French Studies at Louisiana State University in Baton Rouge, Louisiana. He is the author of *Time and Exteriority: Aristotle, Heidegger, Derrida* (Bucknell University Press, 1994), *Political Physics: Deleuze, Derrida and the Body Politic* (Athlone Press, 2001), and co-author, with Marks Bonta, of *Deleuze and Geophilosophy: A Guide and Glossary* (Edinburgh University Press, 2004). In addition, he is co-editor, with Paul Patton, of *Between Deleuze and Derrida* (Continuum Press, 2003) and editor of the *Edinburgh Dictionary of Continental Philosophy* (Edinburgh University Press, 2005; North American edition as *A Dictionary of Continental Philosophy*, Yale University Press, 2006).

James Williams is Professor of European Philosophy at the University of Dundee. He has published widely on contemporary French philosophy. His most recent books include *Understanding Post-structuralism* (Acumen, 2005) and *The Transversal Thought of Gilles Deleuze: Encounters and Influences* (Clinamen, 2006). He is also co-editor, with Keith Crome, of *The Lyotard Reader Guide* (Edinburgh University Press, 2006). He is currently working on a critical reading of Gilles Deleuze's *The Logic of Sense* for Edinburgh University Press.